Embedded System Design

Daniel D. Gajski • Samar Abdi
Andreas Gerstlauer • Gunar Schirner

Embedded System Design

Modeling, Synthesis and Verification

 Springer

Daniel D. Gajski
Center for Embedded Computer Systems
University of California, Irvine
2010, AIR Bldg.
Irvine, CA 92697-2620
USA
gajski@uci.edu

Samar Abdi
Center for Embedded Computer Systems
University of California, Irvine
2010, AIR Bldg.
Irvine, CA 92697-2620
USA
sabdi@uci.edu

Andreas Gerstlauer
Department of Electrical &
Computer Engineering
University of Texas at Austin
1 University Station C0803
Austin, TX 78712
USA
gerstl@ece.utexas.edu

Gunar Schirner
Center for Embedded Computer Systems
University of California, Irvine
2010, AIR Bldg.
Irvine, CA 92697-2620
USA
hschirne@uci.edu

ISBN 978-1-4899-8530-9   ISBN 978-1-4419-0504-8 (eBook)
DOI 10.1007/978-1-4419-0504-8
Springer Dordrecht Heidelberg London New York

Printed on acid-free paper

Springer is part of Springer Science+Business Media (www.springer.com)

Preface

RATIONALE

In the last twenty five years, design technology, and the EDA industry in particular, have been very successful, enjoying an exceptional growth that has been paralleled only by advances in semiconductor fabrication. Since the design problems at the lower levels of abstraction became humanly intractable and time consuming earlier then those at higher abstraction levels, researchers and the industry alike were forced to devote their attention first to problems such as circuit simulation, placement, routing and floorplanning. As these problems become more manageable, CAD tools for logic simulation and synthesis were developed successfully and introduced into the design process. As design complexities have grown and time-to-market have shrunk drastically, both industry and academia have begun to focus on levels of design that are even higher then layout and logic. Since higher levels of abstraction reduce by an order of magnitude the number of objects that a designer needs to consider, they have allowed industry to design and manufacture complex application-oriented integrated circuits in shorter periods of time.

Following in the footsteps of logic synthesis, register-transfer and high-level synthesis have contributed to raising abstraction levels in the design methodology to the processor level. However, they are used for the design of a single custom processor, an application-specific or communication component or an interface component. These components, along with standard processors and memories, are used as components in systems whose design methodology requires even higher levels of abstraction: system level. A system-level design focuses on the specification of the systems in terms of some models of computations using some abstract data types, as well as the transformation or refinement of that specification into a system platform consisting of a set of processor-level components, including generation of custom software and hardware components. To this point, however, in spite of the fact that systems have been manufactured for years, industry and academia have not been sufficiently focused on developing and formalizing a system-level design technology and methodology, even though there was a clear need for it. This need has been magnified by appearance of embedded systems, which can be used anywhere and everywhere, in plains, trains, houses, humans, environment, and manufacturing and in any possible infrastructure. They are application specific and tightly constrained by different requirements emanating from the environment they operate in. Together with ever increasing complexities and market pressures, this makes their design a tremendous challenge and the development of a clear and well-defined system-level design technology unavoidable.

There are two reasons for emphasizing more abstract, system-level methodologies. The first is the fact that high-level abstractions are closer to a designer's usual way of reasoning. It would be difficult to imagine, for example, how a designer could specify, model and communicate a system design by means of a schematic or hundred thousand lines of VHDL or Verilog code. The more complex the design, the more difficult it is for the designer to comprehend its functionality when it is specified on register-transfer level of abstraction. On the other hand, when a system is described with an application-oriented model of computation as a set of processes that operate on abstract data types and communicate results through abstract channels, the designer will find it much easier to specify and verify proper functionality and to evaluate various implementations using different technologies. The second reason is that embedded system are usually defined by the experts in application domain who understand application very well, but have only basic knowledge of design technology and practice. System-level design technology allows them to specify, explore and verify their embedded system products without expert knowledge of system engineering and manufacturing.

It must be acknowledged that research on system design did start many years ago; at the time, however, it remained rather focused to specific domains and communities. For example, the computer architecture community has considered ways of partitioning and mapping computations to different architectures, such as hypercubes, multiprocessors, massively parallel or heterogeneous processors. The software engineering community has been developing methods for specifying and generating software code. The CAD community has focused on system issues such as specification capture, languages, and modeling. However, simulation languages and models are not synthesizable or verifiable for lack of proper design meaning and formalism. That resulted in proliferation of models and modeling styles that are not useful beyond the modeler's team. By introduction of well-defined model semantics, and corresponding model transformations for different design decision, it is possible to generate models automatically. Such models are also synthesizable and verifiable. Furthermore, model automation relieves designers from error-prone model coding and even learning the modeling language. This approach is appealing to application experts since they need to know only the application and experiment with a set of design decisions. Unfortunately, a universally accepted theoretical framework and CAD environments that support system design methodologies based on these concepts are not commercially available yet, although some experimental versions demonstrated several orders of magnitude productivity gain. On the other hand, embedded-system design-technology based on these concepts has matured to the point that a book summarizing the basic ideas and results developed so far will help students and practitioners in embedded system design.

In this book, we have tried to include ideas and results from a wide variety of sources and research projects. However, due to the relative youth of this field, we may have overlooked certain interesting and useful projects; for this we apologize in advance, and hope to hear about those projects so they may be incorporated into future editions. Also, there are several important system-level topics that, for various reasons, we have not been able to cover in detail here, such as testing and design for test. Nevertheless, we believe that a book on embedded system techniques and technology will help upgrade computer science and engineering education toward system-level and toward application oriented embedded systems, stimulate design automation community to move beyond system level simulation and develop system-level synthesis and verification tools and support the new emerging embedded application community to become more innovative and self-sustaining.

AUDIENCE

This book is intended for four different groups within the embedded system community. First, it should be an introductory book for application-product designers and engineers in the field of mechanical, civil, bio-medical, electrical, and environmental, energy, communication, entertainment and other application fields. This book may help them understand and design embedded systems in their application domain without an expert knowledge of system design methods bellow system-level. Second, this book should also appeal to system designers and system managers, who may be interested in embedded system methodology, software-hardware co-design and design process management. They may use this book to create a new system level methodology or to upgrade one existing in their company. Third, this book can also be used by CAD-tool developers, who may want to use some of its concepts in existing or future tools for specification capture, design exploration and system modeling, synthesis and verification. Finally, since the book surveys the basic concepts and principles of system-design techniques and methodologies, including software and hardware, it could be valuable to advanced teachers and academic programs that want to teach software and hardware concepts together instead of in non-related courses. That is particularly needed in today's embedded systems where software and hardware are interchangeable. From this point, the book would also be valuable for an advanced undergraduate or graduate course targeting students who want to specialize in embedded system, design automation and system design and engineering. Since the book covers multiple aspects of system design, it would be very useful reference for any senior project course in which students design a real prototype or for graduate project for system-level tool development.

ORGANIZATION

This book has been organized into eight chapters that can be divided into four parts. Chapter 1 and 2 present the basic issues in embedded system design and discuss various system-design methodologies that can be used in capturing system behavior and refining it into system implementation. Chapter 3 and 4 deal with different models of computations and system modeling at different levels of abstraction as well as system synthesis from those models. Chapter 5, 6, and 7 deal with issues and possible solutions in synthesis and verification of software and hardware component needed in a embedded system platform. Finally, Chapter 8 reviews the key developments and selected current academic and commercial tools in the field of system design, system software and system hardware as well as case study of embedded system environments.

Given an understanding of the basic concepts defined in Chapter 1 and 2, each chapter should be self-contained and can be read independently. We have used the same writing style and organization in each chapter of the book. A typical chapter includes an introductory example, defines the basic concepts, it describes the main problems to be solved. It contains a description of several possible solutions, methods or algorithms to the problems that have been posed, and explains the advantages and disadvantages of each approach. Each chapter also includes relationship to previously published work in the field and discusses some open problems in each topic.

This book could be used in several different courses. One course would be for application experts with only a basic knowledge of computers engineering. It would emphasize application issues, system specification in application oriented models of computation, system modeling and exploration as presented in Chapter 1 - 4. The second course for embedded system designers would emphasize system languages, specification capture, system synthesis and verification with emphasis on Chapter 3, Chapter 4, and Chapter 7. The third course may emphasize system development with component synthesis and tools as described in Chapter 5 - Chapter 8. In which ever it is used, though, we feel that this book will help to fill the vacuum in computer science and engineering curriculum where there is need and demand for emphasis on teaching embedded system design techniques in addition to supporting lower levels of abstraction dealing with circuit, logic and architecture design.

We hope that the material selection and the writing style will approach your expectations; we welcome your suggestions and comments.

Daniel Gajski, Andreas Gerstlauer, Samar Abdi, Gunar Schirner

Acknowledgments

This book was in the making for many years: from concepts to methodologies to experiments. Many generations of researchers at the Center for Embedded Systems at UCI participated in finding and proving what works and what does not. We would like to thank the members of the first generation that established basic principles of embedded systems: Frank Vahid, Sanjiv Narayan, Jie Gong and Smita Bakshi. We would also like to acknowledge the second generation that brought us SpecC and System on Chip Environment: Jianwen Zhu, Rainer Doemer, Lukai Cai, Haobo Yu, Sequin Zhao, Dongwan Shin, and Jerry Peng. And the third generation that made Embedded System Environment available: Lochi Yu, Hansu Cho, Yongyun Hwang, Ines Viskic. In addition, we would like to acknowledge the NISC team: Mehrdad Reshadi, Bita Gorjiara and Jelena Trajkovic for their high-level synthesis contributions and Pramod Chandraria for his work on design drivers.

We would also like to thank Quoc-Viet Dang, who helped us with book formatting, figure creation, generation, and without whom this book would not be possible. We also want to thank our editors Matt Nelson and Brian Thill who made the sentences readable and ideas flow without interruptions. We also want to thank Simone Lacina from grafikdesign-lacina.de for an excellent and artistic cover.

However, the highest credits go to Grace Wu and Melanie Kilian for making our center work flawlessly while we were working and thinking about the book.

Last but not the least, we would like to thank Carl Harris from Springer for encouragement and asking at every conference in the last 5 years the same question: "When is the Orange book coming?"

Contents

List of Figures

List of Tables

Chapter 1

INTRODUCTION

In this chapter we will look at the emergence of system design theory, practice and tools. We will first look into the needs of system-level design and the driving force behind its emergence: increase in design complexity and widening of productivity gap. In order to find an answer to these challenges and find a systematic approach for system design, we must first define design-abstraction levels; this will allow us to talk about design-flow needs on processor and systems levels of abstraction. An efficient design-flow will employ clear and clean semantics in its languages and modeling, which is also, required by synthesis and verification tools. We will then analyze the system-level design flow and define necessary models, define each model separately and its use in the system design flow. We will also discuss the components and tools necessary for system design. We will finish with prediction on future directions in system design and the prospects for system design practice and tools.

1.1 SYSTEM-DESIGN CHALLENGES

Driven by ever-increasing market demands for new applications and by technological advances that allow designers to put complete many-processor systems on a single chip (MPSoCs), system complexities are growing at an almost exponential rate. Together with the challenges inherent in the embedded-system design process with its very tight constraints and market pressures, not the least of which is reliability, we are finding that traditional design methods, in which systems are designed directly at the low hardware or software levels, are fast becoming infeasible. This leads us to the well-known productivity gap generated by the disparity between the rapid paces at which design complexity has increased in comparison to that of design productivity [99].

D.D. Gajski et al., *Embedded System Design: Modeling, Synthesis and Verification,*
DOI: 10.1007/978-1-4419-0504-8_1,
© Springer Science + Business Media, LLC 2009

One of the commonly-accepted solutions for closing the productivity gap as proposed by all major semiconductor roadmaps is to raise the level of abstraction in the design process. In order to achieve the acceptable productivity gains and to bridge the semantic gap between higher abstraction levels and low-level implementations, the goal now is to automate the system-design process as much as possible. We must apply design-automation techniques for modeling, simulation, synthesis, and verification to the system-design process. However, automation is not easy if a system-abstraction level is not well-defined, if components on any particular abstraction level are not well-known, if system-design languages do not have clear semantics, or if the design rules and modeling styles are not clear and simple. In the following chapters, we will show how to answer for those challenges through sound system-design theories, practices, and tools.

On the modeling and simulation side, several approaches exist for the virtual prototyping of complete systems. These approaches are typically based on some variant of C-based description, such as C-based System-Level Design Languages (SLDLs) like SystemC [150] or SpecC [171]. These virtual prototypes can be assembled at various levels of detail and abstraction.

The most common approach in the system design of a many-processor platform is to perform co-simulation of software (SW) and hardware (HW) components. Both standard and application-specific processors are simulated on nstruction-set level with an Instruction Set Simulator (ISS). The custom HW components or Intellectual Property (IP) components are modeled with a timed functional model and integrated together with the processor models into a Transaction-Level Model (TLM) representing the platform communication between components.

In algorithmic-level approaches in designing MPSoCs, we use domain-specific application modeling, which is based on more formalized models of computation, such as process networks or process state machines. These modeling approaches are often supported by graphical capture of models in terms of block diagrams, which hide the details of any underlying internal language. On the other hand, the code can be generated in a specific target language such as C by model-based-design tools from such graphical input.

Such simulation-centric approaches enable the horizontal integration of various components in different application domains. However, approaches for the vertical integration for system synthesis and verification across component or domain boundaries are limited. At best, there are some solutions for the C-based synthesis of single custom hardware units. But no commercial solutions for synthesis and verification at the system level, across hardware and software boundaries, currently exist.

In order to understand system-level possibilities more fully, however, we must step back and explain the different abstraction levels involved in system design.

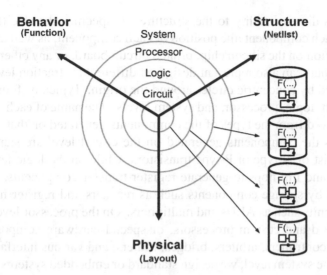

FIGURE 1.1 Y-Chart

1.2 ABSTRACTION LEVELS

The growing capabilities of silicon technology over the previous decades has forced design methodologies and tools to move to higher levels of abstraction. In order to explain the relationship between different design methodologies on different abstraction levels, we will use the Y-Chart, which was developed in 1983 in order to explain differences between different design tools and different design methodologies in which these tools were used [60].

1.2.1 Y-CHART

The Y-Chart makes the assumption that each design, no matter how complex, can be modeled in three basic ways, which emphasize different properties of the same design. As shown in Figure 1.1, the Y-Chart has three axes representing three aspects of every design: behavior (sometimes called functionality or specification), design structure (also called netlist or a block diagram), and physical design (usually called layout or board design). Behavior represents a design as a black box and describes its outputs in terms of its inputs over time. The black-box behavior does not indicate in any way how to build the black box or what its structure is. That is given on the structure axis, where the black box is represented as a set of components and connections. Naturally, the behavior of the black box can be derived from its component behaviors and their connectivity. However, such a derived behavior may be difficult to understand since it is obscured by the details of each component and connection. Physical design adds dimensionality to the structure. It specifies the size (height and width) of each component, the position of each component, as well as each port and connection on the silicon chip, printed circuit board, or any other container.

The Y-Chart can also represent design on different abstraction levels, which are identified by concentric circles around the origin. Typically, four levels are used: circuit, logic, processor, and system levels. The name of each abstraction level is derived from the types of the components generated on that abstraction level. Thus the components generated on the circuit level are standard cells which consist of N-type or P-type transistors, while on the logic level we use logic gates and flip-flops to generate register-transfer components. These are represented by storage components such as registers and register files and by functional units such as ALUs and multipliers. On the processor level, we generate standard and custom processors, or special-hardware components such as memory controllers, arbiters, bridges, routers, and various interface components. On the system level, we design standard or embedded systems consisting of processors, memories, buses, and other processor components.

On each abstraction level, we also need a database of components to be used in building the structure for a given behavior. This process of converting the given behavior into a structure on each abstraction level is called synthesis. Once a structure is defined and verified, we can proceed to the next lower abstraction level by further synthesizing each of the components in the structure. On the other hand, if each component in the database is given with its structure and physical dimensions, we can proceed with physical design, which consists of floorplanning, placement, and routing on the chip or PC board. Thus each component in the database may have up to three different models representing three different axes in the Y-Chart: behavior or function; structure, which contains the components from the lower level of abstraction; and the physical layout of its structure.

Fortunately, all three models for each component are not typically needed most of the time. Most of the methodologies presently in use perform design or synthesis on the system and processor levels, where every system component except standard processors and memories is synthesized to the logic level, before the physical design is performed on the logic level. Therefore, for the top three abstraction levels, we only need a functional model of each component with estimates of the key metrics such as performance, delay, power, cost, size, reliability, testability, etc. Once the design is represented in terms of logic gates and flip-flops, we can use standard cells for each logic component and perform layout placement and routing. On the other hand, some components on the processor-and-system levels may be obtained as IPs and not synthesized. Therefore, their structure and physical design are known, at least partially, on the level higher than logic level. In that case, the physical design then may contain components of different sizes and from different levels of abstraction.

In order to introduce system-level design methodologies we must look first at the design process on each of processor and system abstraction levels.

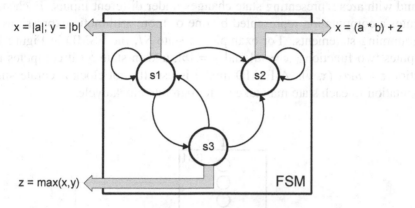

$x = |a|; y = |b|$

$x = (a * b) + z$

$z = \max(x,y)$

FSM

FIGURE 1.2 FSMD model

1.2.2 PROCESSOR-LEVEL BEHAVIORAL MODEL

We design components of different granularity on each abstraction level. On the processor level, we define and design computational components or processing elements (PEs). Each PE can be a dedicated or custom component that computes some specific functions, or it can be a general or standard PE that can compute any function specified in some standard programming language. The functionality or behavior of each PE can be specified in several different ways.

In the early days of computers, their functionality was specified with mathematical expressions or formulas. The functionality of a PE can be also specified with an algorithm in some programming language, or with a flow chart in graphical form. Some simple control functionality, such as controllers or component interfaces, can be specified using the dominant model of computer science, called a Finite State Machine (FSM). A FSM is defined with a set of states and a set of transitions from state to state, which are taken when some input variables reach the required value. Furthermore, each FSM generates some values for output variables in each state or during each transition. A FSM model can be made clock-accurate if each state is considered to take one clock cycle. In general, a FSM model is useful for computations requiring several hundred states at most.

The original FSM model uses binary variables for inputs and outputs. This FSM model can be extended using standard integer or floating-point variables and computing their values in each state or during each transition by a set of arithmetic expressions or programming statements. This way we can extend the FSM model to the model of a Finite State Machine with Data (FSMD) [61]. For example, Figure 1.2 shows a FSMD with three states, *S1*, *S2*, and *S3*, and with arcs representing state changes under different inputs. Each state executes a computation represented by one or more arithmetic expressions or programming statements. For example, in state *S1*, the FSMD in Figure 1.2 computes two functions, $x = |a|$ and $y = |b|$, and in state *S3* it computes the function $z = max\ (x,\ y)$. A FSMD model is usually not clock-accurate since computation in each state may take more than one clock cycle.

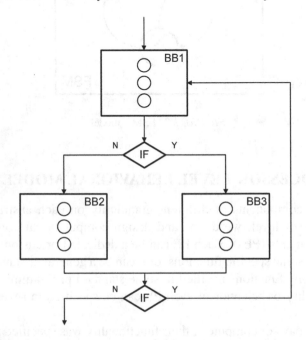

FIGURE 1.3 CDFG model

As mentioned above, a FSMD model is not adequate to represent the computation expressed by standard programming languages such as C. In general, programming languages consist of if statements, loops, and expressions. An if statement has two parts, then and else, in which then is executed if the conditional expression given in the if statement is true, otherwise the else part is executed. In each of the then or else parts, the if statement computes a set of expressions called a Basic Block (BB). The if statement can also be used in the loop construct to represent loop iterations, which are executed as long as the condition in the if statement is true. Therefore, any programming-language code can be represented by a Control-Data Flow Graph (CDFG) consisting of if diamonds, which represent if conditions, and BB blocks, which represent computation [151]. Figure 1.3 shows such a CDFG, this one representing a loop with an if statement inside the loop iteration. In each iteration, the loop construct executes BB1 and BB2 or BB3 depending on the value of the `if` statement. At the end, the loop is exited if all iterations are executed.

A CDFG shows explicitly the control dependencies between loop statements, `if` statements, and BBs, as well as the data dependences among operations inside a BB. It can be converted to a FSMD by assigning a state to each BB and one state for the computation of each `if` conditional. Note that each state in such a FSMD may need several clock cycles to execute its assigned BB or `if` condition. Therefore, a CDFG can be considersd to be a FSMD with superstates, which require multiple clock cycles to execute.

A standard or custom PE can be also described with an Instruction Set (IS) flow chart that describes the fetch, decode, and execute stages of each instruction. A partial IS flow chart is given in Figure 1.4. The fetch stage consists of fetching the new instruction into the Instruction Register (IR) ($IR \leftarrow Mem[PC]$) and incrementing the Program Counter ($PC \leftarrow PC + 1$). In the decode stage, we decode the type and mode of the fetched instruction. In Figure 1.4, there are four types of instructions: register, memory, branch, and miscellaneous instructions. In the case of memory instructions, there are four modes: immediate, direct, relative, and indirect. Each mode contains load and store instructions. Each instruction execution is in turn described by a BB, which may take several clock cycles to execute, depending on the processor implementation structure. For example, the memory-store instruction with indirect addressing computes an Effective Address (EA) by fetching the next instruction pointed to by the PC and uses it to fetch the address of the memory location in which the data will be stored ($EA \leftarrow Mem[Mem[PC]]$). Then it stores the data from the Register File (RF) indicated by the Src1 part of the instruction ($RF[Src1]$) into the memory at location EA ($Mem[EA] \leftarrow RF[Src1]$). Finally, it increments the PC ($PC \leftarrow PC + 1$) and goes to the fetch phase.

The above-described IS flow chart can be converted to a FSMD, where each of the fetch, decode, and execute stages may need one or more states or clock cycles to execute.

In addition to FSMD, CDFG, and IS flow-chart models, other representations can be used to specify the behavior of a PE. They provide differing types of the information needed for the synthesis of PEs. The guideline for choosing one over the other is that more detailed information makes PE synthesis easier.

1.2.3 PROCESSOR-LEVEL STRUCTURAL MODEL

A processor's behavioral model, whether defined by a program in C, CDFG, FSMD, or by an IS, can be implemented with a set of register-transfer components; such a structural model usually consists of a controller and a datapath. A datapath consists of a set of storage elements (such as registers, register files, and memories), a set of functional units (such as ALUs, multipliers, shifters, and

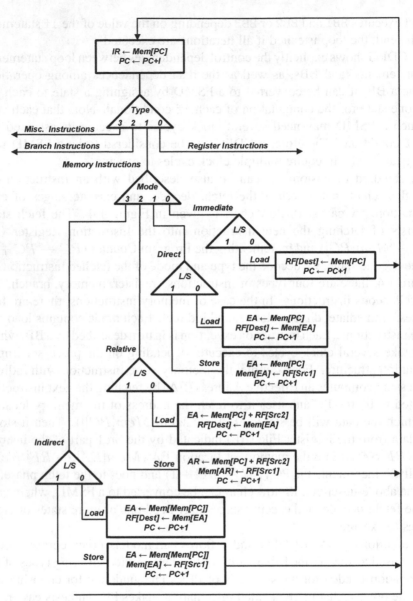

FIGURE 1.4 Instruction-set flow chart

other custom functional units), and a set of busses. All of these register-transfer components may be allocated in different quantities and types and connected arbitrarily through busses or a network-on-chip (NOC). Each component may take one or more clock cycles to execute, each component may be pipelined, and each component may have input or output latches or registers. In addition,

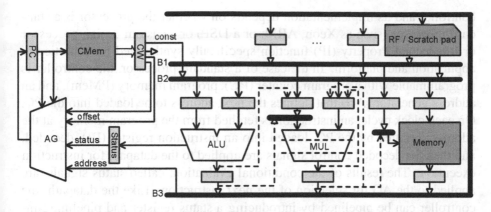

FIGURE 1.5 Processor structural model

the entire datapath can be pipelined in several stages in addition to components being pipelined by themselves. The choice of components and datapath
structure depends on the metrics to be optimized for particular implementation.

An example of such a datapath is shown in Figure 1.5. It consists of a set
of registers and a Register file (RF) or a Scratchpad memory. These storage
elements are connected to the functional units ALU and MUL, and to a Memory
by three busses, B1, B2, and B3. Each of these units has input and output
registers. An ALU can execute an arithmetic or logic operation in one clock
cycle from its input to its output register, while a two-stage pipelined multiplier
MUL needs three clock cycles from its input to its output register. On the other
hand, Memory is not pipelined and requires two clock cycles from its address
register to the output data register. In addition to pipelined functional units such
as the MUL, the whole datapath itself is pipelined. In such pipelined datapath
each operation may take several clock cycles to execute. For example, it takes
three clock cycles from the RF through the ALU input register, the ALU output
register, and back to the RF. On the other hand, it takes five clock cycles through
the MUL, since the MUL is pipelined itself. In order to speed up the execution
for complex expressions such as $a(b+c)$, the datapath allows $(b+c)$ to be sent
directly to the MUL through a data-forwarding path without going back to RF.
In Figure 1.5, such a path is shown going from the ALU output into the left
input register of the MUL. At the same time, this path can also be implemented
by a connection from the ALU output register to the left MUL input. In this
case, we need a short bus, usually implemented with a selector, to select the
ALU output register or the MUL input register as the left inputs to the MUL. A
similar selector is also shown for the Memory-address input, which may come
from the address register or the MUL output register.

The controller defines the state of the processor clock cycle per clock cycle
and issues the control signals for the datapath accordingly. The structure of the

controller and its implementation depends on whether the processor is a standard processor (such as Xeon, ARM, or a DSP) or a custom-design processor or Intellectual Property (IP) function specifically synthesized for a particular application and platform. In the case of a standard processor, the controller is programmable with a program counter (PC), program memory (PMem), and an address generator (AG) that defines the next address to be loaded into the PC. On each clock cycle, an instruction is fetched from the program memory at the address specified by the PC, loaded into an instruction register (IR), decoded, and then the decoded control signals are applied to the datapath for instruction execution. The results of the conditional evaluation, called status signals, are applied to the AG for selection of the next instruction. Like the datapath, the controller can be pipelined by introducing a status register and pipelining instructions from the PC to the IR, through the Datapath and status register and then back to the PC.

In the case of specific IPs or IF components, the controller could be implemented with hardwired logic gates. In terms of digital-design terminology, the PC is then called a State register, the program memory is called output logic, and the AG is called next-state logic.

In the case of specific custom processors, the controller can be implemented with programmability concepts typical of standard processors, and control signal generation of IP implementations. This is shown in Figure 1.5, in which program memory is replaced with control memory (CMem) and instruction register with control word register (CW). CMem stores decoded control words instead of instructions. Figure 1.5 also illustrates how the whole processor is pipelined, including the control and datapath pipelining. On each clock cycle, one control word is fetched from CMem and stored in the CW register. Then the data in the RF are forwarded to a functional unit input register in the next clock cycle, and after one or more clock cycles, the result is stored in the output register and/or in the status register. Finally, in the next clock cycle, the value in the status register is used to select the new address for the PC, while the result from the output register is stored back into the RF or forwarded to another input register.

Selecting components and the structure of a PE and defining register-transfer operations performed in each clock cycle is the task of processor-level synthesis.

1.2.4   PROCESSOR-LEVEL SYNTHESIS

Synthesis of standard processors starts with the instruction set (IS) of the processor. In order to achieve the highest processor performance this process is done manually since standard processors try to achieve the highest performance and minimal power consumption at minimal cost. The second reason for synthesizing processors manually is to minimize the design size and therefore

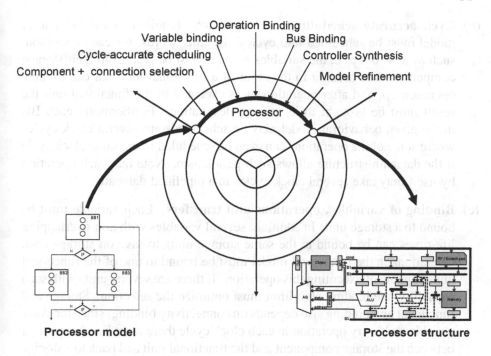

FIGURE 1.6 Processor synthesis

fabrication cost for high-volume production. In contrast, the design or synthesis of a custom processor or a custom IP starts with the C code of an algorithm, which is usually converted to the corresponding CDFG or FSMD model before synthesis and ends up with a custom processor containing the number and type of components connected as required by the given behavioral model. This generation is usually called high-level synthesis or register-transfer synthesis or occasionally just processor synthesis. It consists of five individual tasks.

(a) **Allocation of components and connections**. In processor synthesis, the components are selected from the register-transfer library. It is important to select at least one component for each operation in the behavioral model. Also, it may be necessary to select components that implement some frequently-used functions in the behavioral model. The library must also include a component's characteristics and its metrics, which will be used by the other synthesis tasks. The connectivity among components can be added after binding and scheduling tasks; that way we end up with minimal connectivity. However, we do not know the exact connectivity delays during binding and scheduling. Therefore, it is convenient to also add connections, buses, or a network on a chip, which will allow us to estimate more precisely all the delays.

(b) **Cycle-accurate scheduling**. All operations required in the behavioral model must be scheduled into cycles. In other words, for each operation, such as a = b op c, the variables *b* and *c* must be read from their storage components and brought to the input of a functional unit that can execute operation *op*, and after operation *op* is executed in the functional unit the result must be brought to its storage destination. Furthermore, each BB in the given behavioral model may be scheduled into several clock cycles where some of the operations can even be scheduled in the same clock cycle if the datapath structure allows such parallelism. Note that each operation by itself may take several clock cycles in a pipelined datapath.

(c) **Binding of variables, operations and transfers**. Each variable must be bound to a storage unit. In addition, several variables with non-overlapping life-times can be bound to the same storage units to save on storage cost. Operations in the behavioral model must be bound to one of the functional units capable of executing this operation. If there are several units with such capability, the binding algorithm must optimize the selection. Storage and functional unit binding also depends on connectivity binding, since for every variable and every operation in each clock cycle there must be a connection between the storage component and the functional unit and back to a storage component to which variables and operation are bound.

(d) **Synthesis of controller**. The controller can be programmable with a read-write program memory or just a read-only memory for fixed-functionality IPs. The controller can be also implemented with logic gates for small control functions. As mentioned earlier, the program memory can store instructions or just control words which may be longer then instructions but require no decoding.

(e) **Model refinement**. A new processor model can be generated in several different styles with complete, partial, or no binding. For example, the statement a = b + c executing in state *(n)* can be written:

   (1) without any binding:
        a = b + c;

   (2) with storage binding of *a* to RF(1), *b* to RF(3), and *c* to RF(4):
        RF(1) = RF(3) + RF(4);

   (3) with storage and functional unit binding with + bound to ALU1:
        RF(1) = ALU1(+,RF(3),RF(4));

   (4) or with storage, functional unit, and connectivity binding:
        Bus1 = RF(3); Bus2 = RF(4); Bus3 = ALU1
        (+,Bus1,Bus2); RF(1) = Bus3;

A structural model can be also written as a netlist of register-transfer components, in which each component is defined by its behavior from the component library.

Tasks (a), (b), and (c) can be performed together or in any specific order, but they are interdependent. If they are performed together, the synthesis algorithm becomes very complex and unpredictable. One strategy is to perform allocation first, followed by binding and then scheduling. Another possibility is to do a complete allocation first, followed by storage binding, while combining unit and connectivity binding with scheduling.

Any of the above tasks can be performed manually or automatically. If they are all done automatically, we call the above process processor-level or high-level synthesis. On the other hand, if (a) to (d) are performed manually and only (e) is done automatically, we call the process model refinement. Obviously, many other strategies are possible, as demonstrated by the number of design-automation tools available that perform some of the above tasks automatically and leave the rest for the designer to complete.

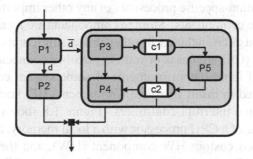

FIGURE 1.7 System behavioral model

1.2.5 SYSTEM-LEVEL BEHAVIORAL MODEL

Processor-level behavioral models such as the CDFG can be used for specifying a single processor, but will not suffice for describing a complete system that consist of many communicating processors. A system-level model must represent multiple processes running in parallel in SW and HW. The easiest way to do this is to use a model which retains the concept of states and transitions as in a FSM but which extends the computation in each state to include processes or procedures written in a programming language such as C/C++. Furthermore, in order to represent a many-processor platform working in parallel or in pipelined mode, we must introduce concurrency and pipelining. Since processes in a system run concurrently, we need a synchronization mechanism for data exchange, such as the concept of a channel, to encapsulate data communication. Also, we need a model which supports hierarchy, so as to allow designers to write complex system specifications without difficulty. Figure 1.7 illustrates such a model of hierarchical sequential-parallel processes, which is usually called a Process State Machine (PSM). This particular PSM is a system-level behavior or system specification, consisting of processes *P1* to *P5*. The system starts with *P1*, which in turn triggers process *P2* if condition *d* is true, or another process consisting of *P3*, *P4*, and *P5* if condition *d* is not true. *P3* and *P4* run sequentially and in parallel with *P5*, as indicated by the vertical dashed line. When either *P2* is finished or the sequential-parallel composition of *P3*, *P4*, and *P5* is finished, the execution ends.

1.2.6   SYSTEM STRUCTURAL MODEL

A system-level structural model is a block diagram or a netlist of system components used for computation, storage, and communication. Processing Elements (PEs) can be standard processors or custom-made processors. They can also be application-specific processors or any other imported IPs or special-functions hardware components. Storage components are local or shared memories which may also be included in other processing components. Communication Elements (CE) are buses or routers possibly connected in a Network-on-Chip (NOC). If input-output protocols of some system component do not match, we will need to insert Interface Components (IF) such as transducers, bridges, arbiters, and interrupt controllers. Figure 1.8 shows a simple system platform consisting of a CPU processor with a local memory, an IP component, a specially-designed custom HW component (HW), and the shared memory (Mem). They are all connected through two buses, the CPU bus and IP bus. Since CPU and IP buses use different protocols, a special IF unit (Bridge) is included. The HW unit has the IF for the CPU bus protocol already built into it. Since the CPU bus has CPU and HW components competing for the bus, a special IF component (Arbiter) is added to grant bus access to one of the requesting components.

A system structural model is generated from the given behavioral model by the process called system synthesis.

1.2.7   SYSTEM SYNTHESIS

System synthesis starts with system-level behavioral model, such as the one shown in Figure 1.7, and generates the system structure, which consists of standard or custom PEs, CEs, and SW/HW IF components, as shown in Figure 1.8. Standard components, including their functionality and structure, can be found in the system-level component library, while custom components must be de-

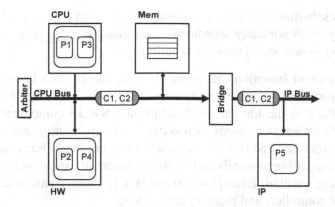

FIGURE 1.8 System structural model

fined and synthesized on the processor level before they can be included in the library. According to the given definition, the behavioral model is a usually a composition of two objects: processes and channels. The structural model, on the other hand, uses different objects: processes are executed by PEs such as standard processors, custom processors, and IPs, and channels are implemented by buses or NoCs with well-defined protocols. The behavioral model can be converted into an optimized system platform by the following set of tasks, as shown in Figure 1.9:

(a) **Profiling and estimation.** Synthesis starts by profiling the application code in each process and collecting statistics about types and frequency of operations, bus transfers, function calls, memory accesses, and about other statistics that are then used to estimate design metrics for the optimization of the platform or application code. These estimated metrics include performance, cost, bus traffic, power consumption, memory sizes, security, reliability, fault tolerance, among others;

(b) **Component and connection allocation.** Next, components from the library of standard and custom processors, memories, IPs, and custom-functionality components must be allocated and connected with buses through bridges or routers. It is also possible to start with a completely defined platform, which is very useful for application and system software upgrades and product versioning;

(c) **Process and channel binding.** Processes are assigned to PEs, variables to memories (local and global), and channels to busses. This requires an optimized partitioning of processes, variables, and connection traffic to minimize the platform-design metrics;

(d) **Process scheduling**. Parallel processes running on the same PE must be statically or dynamically scheduled. This requires generating a real-time operating system for dynamic scheduling;

(e) **IF component insertion**. Required IF components must be inserted into the platform from a library or synthesized on the processor level before being added to the library. Such additional SW IF components include system firmware components such as device drivers, routing, messaging and interrupt routines, and HW IF components to connect platform components with incompatible protocols and facilitate communication synchronization or message queuing. Examples of these HW IF components would include interrupt controllers and memory controllers.

(f) **Model refinement**. The final step in converting a behavioral model into an optimized system platform consists of refining the behavioral model into a structural model in order to reflect all the platform decisions, as well as adding newly synthesized SW, HW, and IF components.

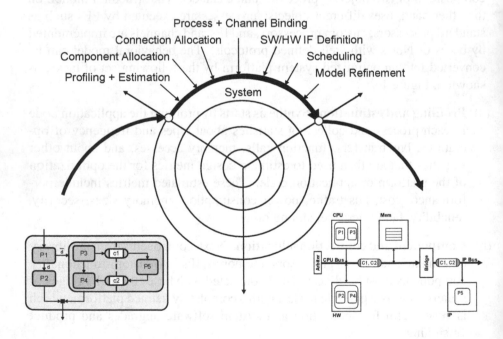

FIGURE 1.9 System synthesis

The above tasks can be performed automatically or manually. Tasks (b)-(e) are usually performed by designers, while tasks (a) and (f) are better done automatically since they require too many pain-staking and error-prone statistical accounting or code construction. Once the refinement is performed, the structural model can be validated by simulation quite efficiently since all the component behaviors are described by high-level functional models. More formal verification of the behavioral and structural models is also possible if we formalize the refinement rules.

In order to generate a cycle-accurate model, however, we must replace each functional model of each component with a cycle-accurate structural model for custom HW or IS model for standard processors executing compiled application code. Once we have this model, we can refine it further into a cycle-accurate model by performing RTL synthesis for custom processors or custom IFs, and by compiling the processes assigned to standard processors to the instruction-set level and inserting an IS simulator to execute the compiled instruction stream. We also have to synthesize system software or firmware for the standard and custom processors. After RTL/IS refinement, we end up with a cycle-accurate model of the entire system. This model can be downloaded to a FPGA board by using standard CAD tools provided by the board supplier. This way we can obtain a system prototype. If all synthesis and refinement tasks are automated, the system prototype can be generated in a few weeks, depending on the expertise of the system and application designers.

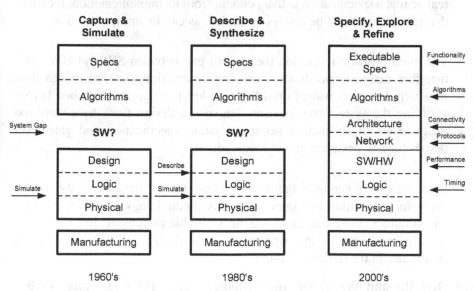

FIGURE 1.10 Evolution of design flow over the past 50 years

1.3   SYSTEM DESIGN METHODOLOGY

Design flow has been changing with the increase in system complexity over the past half-century. We can indicate several periods which resulted in drastic changes in design flow, tools, and methodology, as shown in Figure 1.10.

(a) **Capture-and-Simulate methodology (1960s to 1980s)**. In this methodology, software and hardware design was separated by a so-called system gap. SW designers tested some algorithms and occasionally wrote the requirements document and the initial specification. This specification was given to the HW designers, who began the system design with a block diagram based off of it. They did not, however, know whether their design would satisfy the specification until the gate-level design was produced. When the gate netlist was captured and simulated, designers could determine whether the system worked as specified. Usually this was not the case, and therefore the specification was usually changed to accommodate implementation capabilities. This approach started the myth that specification is never complete. It took many years for designers to realize that a specification is independent from its implementation, meaning that specification can be always upgraded, as can its implementation.

The main obstacle to closing the system gap between SW and HW , and therefore between specification and implementation, was the design flow in which designers waited until the gate level design was finished before verifying the system specification. In such a design flow there were too many levels of abstraction between system specification and gate level design for SW designers to get involved.

Since designers captured the design description at the end of the design cycle for simulation purposes only, this methodology is called capture-and-simulate. Note that there was no verifiable documentation before the captured gate level design, since most of the design decisions were stored informally in the designers' minds.

(b) **Describe-and-Synthesize methodology (late 1980s to late 1990s)**. The 1980s brought us tools for logic synthesis which have significantly altered design flow, since the behavior and structure of a design were both captured on the logic level. Designers specified first what they wanted in Boolean equations or FSM descriptions, and then the synthesis tools generated the implementation in terms of a logic-level netlists. In this methodology therefore, the behavior or function comes first, and the structure or implementation comes afterwards. Moreover, both of these descriptions are simulatable, which is an marked improvement over Capture-and-Simulate methodology, because it permits much more efficient verification; it makes it possible to verify the descriptions' equivalence since both descriptions can in principle be reduced to a canonical form. However, today's designs are too large for this kind of equivalence checking.

By the late 1990s, the logic level had been abstracted to the Register-Transfer Level (RTL) with the introduction of cycle-accurate modeling and synthesis. Therefore, we now have two abstraction levels (RTL and logic levels) and two different models on each level (behavioral and structural). However, the system gap still persists because there was not relation between RTL and higher system level.

(c) **Specify, Explore-and-Refine methodology (early 2000s to present)**. In order to close this gap, we must increase the level of abstraction from the RTL to the system level (SL) and to introduce a methodology that includes both SW and HW. On the SL, we can start with an executable specification that represents the system behavior; we can then extend the system-level methodology to include several models with different details that correspond to different design decisions. Each model is used to prove some system property: functionality, application algorithms, connectivity, communication, synchronization, coherence, routing, performance, or some design metric such as performance, power, and so on. So we must deal with several models in order to verify the impact of design decisions on every metric starting from an executable specification down to the RTL and further to the physical design. We can consider each model as a specification for the next level model, in which more implementation detail is added after more design decisions are made. We can label this a Specify-Explore-Refine (SER) methodology [63, 100], in that it consists of a sequence of models in which each model is a refinement of the previous. Thus SER methodology follows the natural design process in which designers specify the intent first, then explore possibilities, and finally refine the model according to their decisions. SER flow can therefore be viewed as several iterations of the basic Describe-and-Synthesize methodology.

In order to define a reasonable SER methodology, we need to overview the status of methodologies presently in use, their shortcomings, and how to upgrade them to the system level. More detailed explanations will be given in Chapter 2.

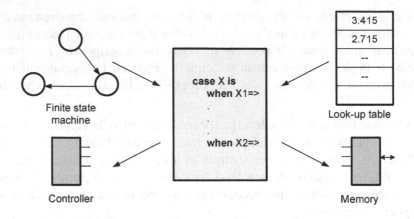

FIGURE 1.11 Missing semantics

1.3.1   MISSING SEMANTICS

With the introduction of system-level abstraction, designers must generate even more models. One obvious solution is to automatically refine one model into another. However, that requires well-defined model semantics, or, in other words, a good understanding what a given model means. This is not as simple as it sounds, since design methodologies and the EDA industry have been dominated by simulation-based methodologies in the past. For example, the models written in Hardware Description Languages (HDLs) (such as Verilog, VHDL, SystemC, and others) are simulateble, but they are not really synthesizable or verifiable. They can result in ambiguities that make automated synthesis and verification impossible, due to the unclear semantics involved. Only a well-defined subset of these languages may be synthesizable or verifiable.

As an example of this problem, we can look in Figure 1.11 at a simple case statement available in any hardware or system modeling language. This type of case statement can be used to model a FSM in which every case such as *X1*, *X2*, ..., represents a state in which all its next states are defined. This type of case statement can also be used to model a look-up table, in which every case *X1, X2*, ..., indicates a location in the memory that contains a value in the table. Therefore, we can use the same case statement with the same variables and format to describe two completely different components. Unfortunately, FSMs and look-up tables require completely different implementations: a FSM can be implemented with a controller or with logic gates, while a look-up table is usually implemented with some kind of memory. It is also possible to implement a FSM with a memory or a table using logic gates. However, this would not be a very efficient implementation, and it would not be acceptable to any designer. So a model which uses case statements to model FSMs and tables is good for simulation but not for implementation because neither a designer nor a synthesis tool can determine which type of structure was described by the case statement.

The lesson is that contemporary modeling languages allow modelers to describe the design in many different ways and to use the same description for different designs details. But for automatic refinement, synthesis, and verification, we need clean and unambiguous semantic which uniquely represents all the system concepts in a given model. Such a clean semantic is missing from most of the simulation-oriented languages. In order to have well-defined semantics, we need to introduce some form of formalism to models and modeling languages.

1.3.2 MODEL ALGEBRA

We can see than that in order to find an acceptable methodology and develop adequate system design tools, we need to clearly define the semantics of the different design models and the rules for model refinement. Generally speaking, every model is a set of objects and composition rules. In order to find an adequate model structure, we can look to some standard, well-defined algorithmic structures such as arithmetic algebra, which consists of objects such as numbers and operations such as addition and multiplication; we can represent this as, numbering here

$$Algebra :< objects, operations >$$

Algebra's composition rules allow for the creation of hierarchical expressions such as $a * (b + c)$ and their transformation. For example, by multiplying a with the $(b + c)$ we get a new expression numbering here

$$a * (b + c) \quad = \quad a * b + a * c$$

The equivalence of these two expressions allows designers to perform optimization using arithmetic algebra rules. The expression on the left requires one multiplier and one adder and may take two clock cycles, assuming each operation takes one clock cycle. On the other hand, the expression on the right requires two multipliers and one adder; it may also take two clock cycles to complete. In this case, the expression on the left requires fewer resources for the same execution time. In case we are limited to only one adder and one multiplier, the expression on the right would take three clock cycles. In this case, the expression on the left would execute faster for the given resources. Thus, arithmetic algebra allows for the creation of expressions and their transformation to equivalent expressions for the optimization of some design metrics.

We can also create model algebra consisting of modeling objects and composition rules similar to those of arithmetic algebra, in that numbering here

$$Model Algebra :< objects, compositions >$$

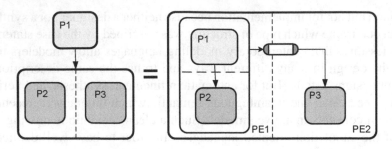

FIGURE 1.12   Model equivalence

The most important objects for a model are processes or behaviors and com-munication channels. Using model algebra composition rules, objects can be composed hierarchically in sequential and concurrent fashion. The left side of Figure 1.12 for example shows a sequential composition of process *P1* fol-lowed by a concurrent composition of processes *P2* and *P3*. After deciding that the system platform will have two processing elements, *PE1* and *PE2*, and mapping *P1* and *P2* to *PE1*, and *P3* to *PE2*, the original model must be refined to reflect these platform and mapping decisions. The necessary transforma-tions include the introduction of another level into the hierarchy to reflect the given platform architecture and a new communication channel for data transfer. The new channel, in addition to transferring data, preserves the sequentiality of *P1* and *P3*, since data is transferred after *P1* finished and *P3* can not start before data becomes available. Thus the model on the right side is equivalent to the model on the left side. Thus, model algebra allows creation of models and proving their equivalence by allowing model transformations that preserve their execution semantics. In other words, model algebra is an enabling technology for system synthesis and verification.

Model algebra also enables the development of a Specify-Explore-Refine (SER) methodology, as shown in Figure 1.13. With model algebra in place, we can define the semantic and style of each model and define a model transforma-tion or refinement for each design decision. Therefore, after some estimation and exploration, a design decision can be made which will in turn result in a model transformation that preserves execution equivalence. A model transfor-mation usually results in the replacement of one object by several other objects or in a re-composition of some objects. In case of system-level design, as shown in Figure 1.13, we start with the system executable specification, generate sev-eral intermediate models, and end up with a cycle-accurate model that can be downloaded to a FPGA board with standard FPGA tools or synthesized into an ASIC with adequate ASIC tools.

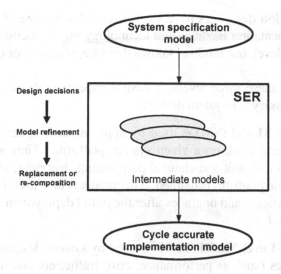

FIGURE 1.13 SER Methodology

1.4 SYSTEM-LEVEL MODELS

In the most common industrial-system design flow today, designers are using different modeling styles and different numbers of models to demonstrate the validity of their software or hardware. The key question facing a designer is how many models are really needed. At one extreme is the argument that we need one model for each level of abstraction and each design metric. Since different metrics are used by different design groups, this strategy will result in many incompatible models in the same organization and will eventually break down the product design flow. At the other extreme, some designers claim that one model is good enough for several abstraction levels and many metrics. This strategy generates a very complex and overly-detailed model, while degrading its comprehensibility and, therefore decreasing design productivity.

In order to identify the minimum number of models necessary for system design, we must look into the profiles and expertise of the designers of such systems. There are three types of designers: application, system, and implementation designers.

Application designers have a good knowledge of their application domain, application structures, and application algorithms, but only a basic knowledge of system design and technology;

System designers have a good knowledge of system organizations, multiprocessor architectures and their operations, and system-level SW and HW, but overview knowledge of application and implementation technology;

Implementation designers have a specialized knowledge of specific components, implementation methods and technology on abstraction levels below than the system level, but minimal knowledge of application or overall system operation.

According to these three levels of design expertise, at least three system models are necessary for system design:

(a) **Specification Model (SM)** is used by application designers to prove that their algorithms work on a given system platform. They also modify it for more efficient task and channel partitioning, mapping, and scheduling once a possible platform is defined. This model is also used for adding new features, functions, and upgrades after the initial deployment of the product into the market.

(b) **Transaction-Level Model (TLM)** is used by system designers to estimate design metrics (such as performance, cost, traffic, communication, power consumption, reliability, among others) and to explore different implementations (component selection, connectivity, system firmware, operating-system selection, and different types of interfaces).

(c) **Cycle-Accurate Model (CAM)** is used by HW designers to verify the correctness of the generated system HW components (such as custom processors, interfaces, memory controllers, interrupt controllers, bridges, arbiters, and similar components) and by SW designers to verify the system firmware (such as task scheduling, data transfer, synchronization, interrupt procedures, and routing).

Given the ongoing advances in technology and systems, the distinction between HW and SW designers is blurring; every embedded system includes both hardware and software, and furthermore, what is in hardware today could be moved to software in the next version of the product and vice versa. Any of above three essential models can be used by either of the application, SW or HW groups depending on the company's organization. Since the above models are defined according to designers' expertise, we may want to explore what each of these models contains and how it is used.

The SM contains the application code and the system requirements, as shown in Figure 1.7, in which the application code is defined by a PSM model of computation. This model is used to specify application algorithms and to optimize the application code for mapping to a platform when the platform is defined. In other words, the application code must be broken down into processes communicating through channels in a way that will optimize local communication and minimize long-distance communication after platform mapping. Similarly, the application code must be divided into processes whose computation structure will match the PE types in the platform so as to facilitate performance optimization. In the same way, the application code can be restructured to match other design metrics. This code restructuring may take several iterations.

Platform architecture consists of a set of PEs and a set of CEs selected from the library or defined by the user, as shown in Figure 1.8. The platform architecture can be given with the SM or defined partially or completely during the system synthesis process shown in Figure 1.9. More components or connections can be added or modified at a later stage for the optimization of some metrics. During synthesis, processes are mapped to the architecture's PEs, that is, its CPUs, HW components, and IPs. Variables are mapped to memory components, either local to PEs or shared. Channels between processes are mapped to routes consisting of buses and bridges, as described in the previous sections.

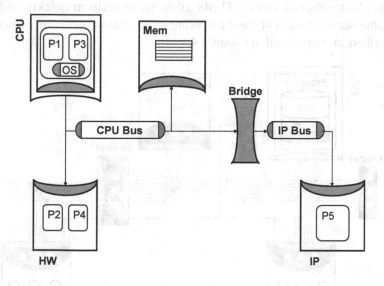

FIGURE 1.14 System TLM

The given or generated platform with its mapped application processes and channels defines the system structure after system synthesis, as shown previously in Figure 1.9. This system structure is usually modeled with a TLM, which Figure 1.14 illustrates. This TLM has the same components as Figure 1.9 but with added bus interface functions and with channels combined into buses. Channels *C1* and *C2* use the CPU bus and the IP bus through the bridge, which converts the CPU bus protocol into IP bus protocol. On the system software side, the CPU has the operating system (OS) model added, as well as communication drivers for the CPU bus.

The TLM can be un-timed or timed. For the timed TLM, we need to profile the application code and generate performance estimates for each process on its corresponding component, in addition to estimating the time needed to send each message on its corresponding route. For example, for a message sent from

P3 to P5, we must add time estimates for the message transfer over the CPU bus, protocol conversion in the bridge and the transfer over the IP bus.

As mentioned earlier, a TLM is used to explore different SM structures, different platform architectures, each with different numbers and types of components and different SM mappings to platform architectures. For example, we may want to restructure a SM for a given platform and a given mapping. On the other hand, for any given SM and component library, we may try to generate an optimal platform for some given metric. Similarly, for a given SM and platform, we may try to find an optimal mapping for a given metric. In an extreme case, we may want to simultaneously optimize the SM structure or application code, platform architecture, and the mapping for one or several metrics. As mentioned earlier, TLMs allow us to perform quickly and with reasonable accuracy each of these disparate optimization scenarios, which will be described in more detail in Chapter 4.

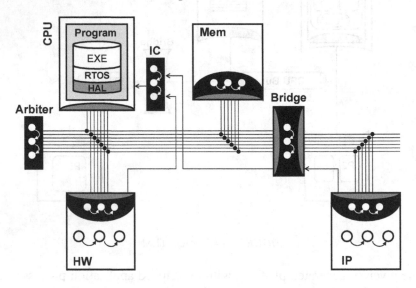

FIGURE 1.15 System CAM

A TLM serves for exploration of the platform structure and estimation of system quality metrics. It is a system-level model. In order to generate a prototype or a product we need to lower the abstraction level to the processor level and generate a CAM of the entire system, example of which is shown in Figure 1.15. From a TLM model, we can generate the CAM by refining the functionality of each component to the cycle-accurate or register-transfer level (RTL) description. For custom HW components we need to generate RTL using processor-level synthesis that has been described in Figure 1.6. For IF components such as arbiters, interrupt controllers, memory controllers, and CD components, such as bridges and routers, we can use the same processor-level synthesis. In case of IP components we can replace functional description with the RTL description that is provided by the IP suppliers. In case of standard processors we can compile application and system code into an instruction stream to be executed on the processor's RTL model or IS model. In case of system prototyping, the instruction stream is executed on the processor available on the prototyping board. CAM model also includes models of system SW including communication drivers, libraries, and RTOS.

Depending on the prototype target, the CAM must include all files that will be needed by the respective FPGA board design tools. These files and the CAM model are exported to the FPGA design environment where implementation designers compile the SW and perform synthesis for the HW components. Finally, a bit-stream is generated that directly programs the FPGA with the prototype. This programmed FPGA typically has a hyper-terminal user interface that can be used to debug the prototype.

In a different scenario, a CAM can co-simulate with an instruction-set simulator and a hardware-description language simulator. In this case, an ISS is inserted and all the SW code is compiled into binaries and simulated with the inserted ISS while the hardware is simulated on the RTL level in Verilog or VHDL with an appropriate co-simulator.

Each of these different model types, SM, TLM, and CAM, has a place in every system design-flow. Their roles in different design methodologies will be elaborated in Chapter 2 and Chapter 3.

1.5   PLATFORM DESIGN

Today's platforms come in a variety of forms and shapes. Usually they have one or more standard processors, a processor for multi-media, several different specialty IPs, and a multitude of special IFs for different communication standards. Although platforms today use a variety of components, platform structure is not standardized. System-design automation is possible if we can limit the number of components and the structure of each platform.

In order to do that we can identify a small number of necessary component types:

(a) **processing components** (PEs), such as standard or custom processors, for computation tasks;

(b) **storage components**, such as local and global memories, for data storage;

(c) **communication components** (CEs), such as transducers and bridges, for the communication of data, temporary buffering of data, and translation of one communication protocol into another;

(d) **interface components** (IFs), such as arbiters for bus traffic regulation, DMA controllers for speeding up memory traffic, interrupt controllers for synchronization, UARTs, and others.

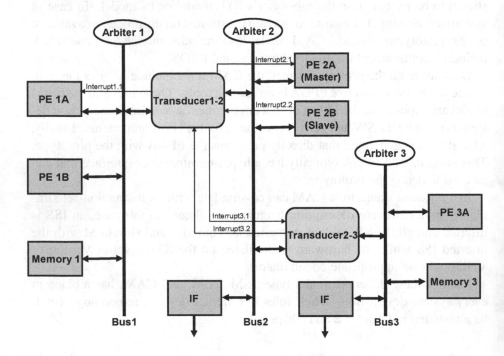

FIGURE 1.16 Platform architecture

These four component types could be sufficient for any platform, although some others (analog components, sensors) may be added for specific application functions. Basic connectivity can be accomplished with buses and transducers to convert one protocol into another if necessary. Transducers can also be used for routing if they can compute possible routes or if the routing is encoded in the message. Such transducers are sometimes called routers; they can be combined into a variety of networks-on-chips (NOCs).

Using components *(a)* through *(d)* we can construct any system platform. Figure 1.16 gives an example of such a platform. It consists of three buses, *Bus1*, *Bus2*, and *Bus3*, with an arbiter on every bus. *Bus1* has two PEs, *PE 1A* and *PE 1B*, and one memory. *Bus2* has also two PEs, *PE 2A* and *PE 2B*, where *PE 2A* is the bus muster. *Bus3* on the other hand has one PE and one memory. There are two transducers: *Transducer 1-2* between *Bus1* and *Bus2*, and *Transducer 2-3* between *Bus2* and *Bus3*. In order for *PE 1A* to send a message to *PE 3A*, the message must be routed through both transducers.

We can build any platform structure using PEs as computation components, memories as storage components, and buses as connectivity components. We can build more complex system connectivity structures by using transducers as routers and building complex NoCs with such transducers. Finally, we conclude that four component types *(a)* through *(d)* are sufficient to build a platform of any complexity. The simplicity of the platform construction rules as demonstrated in Figure 1.16 enables system design automation for modeling, synthesis, and verification.

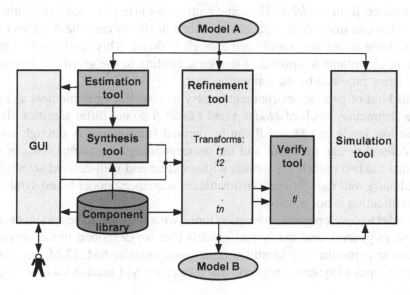

FIGURE 1.17 General system environment

1.6 SYSTEM DESIGN TOOLS

In order to explore system design automation, we must look at basic system design needs and some possible generic system tools. Using a Specify-Explore-Refine methodology, system models can be generated automatically as long as for every design decision we can define a sequence of transformations that will refine the given model accordingly. Therefore, for every pair of models (such as *Model A* and *Model B* in Figure 1.17), we may make several design decisions at once and apply a transformation sequences to refine the model. The tool that performs that refinement is called the refinement tool. Obviously, we can verify that the applied transformations will preserve execution equivalence using a verify tool that will verify each sequence of transformations. The new *Model B* can be also simulated and the results compared with the results from *Model A* by way of a simulation tool.

The design decisions which guide the refinement are made by the designer through a GUI, which also includes selecting the proper components for the design refinement from a component library. The refinement tool may also select components during the refinement transformations when a functional description is replaced with a structural description. In order for a designer to make correct design decisions, an estimation tool provides the values of different metrics obtained from *Model A*. The same metrics are also provided to a synthesis tool which can make design decisions automatically in case the designers are not available or are not experienced enough to do so. This synthesis tool uses different algorithms to optimize a design according to the given requirements and metrics provided by the estimation tool.

This kind of general environment is easy to visualize or implement as long as the abstraction levels of *Model A* and *Model B* do not differ significantly in abstraction levels and *Model B* can be derived from model A through set of well-defined design decisions and the corresponding transformations. In the case that the two models are written without clear and well-defined semantics, or without a well thought-out transformation sequence, model-based synthesis and verification is not possible.

In order to develop good automation tools, we must look at the system design process in general, and the system models that we described in the previous sections in particular. We identified three basic models: SM, TLM, and CAM for three types of system designers. Therefore, we will need at least two types of tools:

(a) a front-end tool, which is a tool for application developers who want to test their product concepts. It captures system behavior with a MoC in a standard language such as C, C++, SystemC, Matlab, UML or a similar graphic representation as an input, and it generates a functional or a timed TLM to use in design-space exploration;

(b) a back-end tool, which is a tool for SW and HW system and implementation designers to use in creating the SW and HW details for a given platform and for a particular application. Such tool may take the TLM model from the front-end and generate a CAM or PCAM in a HDL for the HW and an IS model for the SW, to be used for co-simulation with a HDL/IS co-simulator. The HW model can be synthesized with RTL tools while the compiled SW instruction stream can be downloaded to selected processors in the prototype or the final implementation of the system platform.

This sort of generic system-development tool is shown in Figure 1.18. It consists of a Front-End stage and a Back-End stage, which are supported by two types of interfaces.

The Front-End consists of two elements, System Capture and Platform Development. System capture may be a graphical user interface which captures

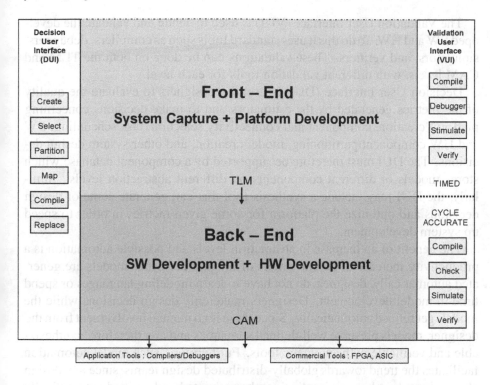

FIGURE 1.18 System tools

the definition of the platform architecture and the product application code with a SM. The Platform Development tool generates timed TLMs of the platform architecture which executes the product application captured by the capture tool. In order to generate a timed TLM, the Front-End tool uses an estimator for estimating computation time for each PE and communication time for each message on a particular message route. Such timed TLMs provide reliable metrics for early exploration of design choices. In addition, TLMmay provide estimates of other metrics, such as power, cost, and reliability. If these metrics are not satisfactory, a designer may change the type and number of components, the number and type of connections, and/or the mapping of the application code onto different PEs and CEs. The TLM generated by the Platform Development tool models the platform structure and HW, as well as SW components such as RTOS.

In the Back-End, the HW Development component is used to generate a cycle-accurate or RTL description of the HW components, which can be further refined by commercially-available tools for ASIC or FPGA manufacturing. SW Development generates the firmware necessary to run communication and application SW on the platform.

The Validation User Interface (VUI) is used to debug and validate the developed SW and HW. To do this it uses standard tools such as compilers, debuggers, simulators, and verifiers. These validations can be done on both the TLM and CAM levels, with different validation tools for each level.

Decision User Interface (DUI) is used by designers to evaluate the quality of the metrics generated by the estimators and to make decisions concerning platform creation, component and connectivity selection, task scheduling, SW and HW component partitioning, model creation, and other system design decisions. The DUI must therefore be supported by a component database which stores models of different components on different abstraction levels. Similarly, the DUI may include a synthesis tool that can generate some of design decisions and optimize the platform for some given metrics in order to speed up system development.

The benefit of an increase in abstraction levels and possible automation is a productivity increase of several orders of magnitude. Since models are generated automatically, designers do not have to learn modeling languages or spend time in model development. Designers make only design decisions while the models get refined automatically. Since there is no manual involvement from the designer, models preserve well-defined semantics and are, therefore, synthesizable and verifiable with automatic tools. Furthermore, such model automation facilitates the trend towards globally-distributed design teams, since any design change introduced at any location can be estimated and verified automatically at any other location. It also enables cooperation between component suppliers and product integrators, as, for instance, in the automotive market, where many suppliers provide hundreds of electronic control units to car manufacturers for integration into a vehicle; model automation enables efficient negotiation and coordinated discussion regarding integrators requirements and component supplier offerings. In addition to the productivity benefits, the main benefit of model automation is that it does not require high-level design expertise, which allows many application experts without detailed knowledge of embedded system design to develop systems and upgrades for their products. This also speeds up the deployment of products to the market since long redesign processes can be avoided.

1.7   SUMMARY

The concepts presented so far seem reasonable but the main question still remains: "Do they work?" Let's look at them one at the time.

We described a methodology that is based on well-defined models, design decisions, and model transformations. Simple models and clear semantics worked in the past as an enabler for the progress to the next level of abstraction.

For example, in 1960s, we had many different logic design styles, from resistor-transistor logic (RTL), to diode-transistor logic (DTL), to transistor-transistor logic (TTL), and so on. Real progress in design productivity only came when we reduced the number of components on the chip to two types of transistors, P-type and N-type, and incorporated them into CMOS logic, which is still in use today.

Another example of simple and clear semantics comes from the experiences in layout design. The real progress in layout floorplanning, placement and routing algorithms, and layout tools was made when we reduced logic gates to standard cells and routing to channel routing. Again, this simplification made layout tools more efficient and acceptable by designers. Logic synthesis provides another example of efficiency introduced by a clear and simple semantics; here the real progress was made when number of components was reduced to simple NAND, NOR, and NOT gates. It also worked on RTL level, when the controller design was reduced to the FSM model for a datapath structure that was manually specified by the designer. Unfortunately, the definition of a simple and clear semantics for the processor-level design, consisting of a custom controller and a custom datapath, is still not available. There are many C-to-RTL synthesis tools based on FSMD models with a variety of synthesis algorithms providing results of varying quality. This indicates that there is still a need for a clarifying and simplifying high-level synthesis methodology and the accompanying tools. On the next abstraction level, the system level automation is still in its infancy. In order to automate system design, which would be the ideal, we needed to simplify even more the components, models, and tools involved. For that reason, we described a possible strategy called model algebra to properly define models and refinements.

As proof for the success of this approach, we point to several academic and commercial system tools that have been developed for automatic model generation, synthesis, and verification. The preliminary evidence testifies that these tools result in productivity improvements of several orders of magnitude. In addition, the clear semantics we advocate allow easy management of globally-distributed projects, since upgrades made anywhere in the world can be checked and verified everywhere. They also allows for easy product versioning and the timely creation of product derivatives, which will result in shorter time-to-market and more profitability.

As mentioned above, the complexities of embedded systems are forcing the design and design automation communities to rethink the design flow, modeling, synthesis, and verification of such systems. In addition, these complexities are also requiring a more substantial and deeper change in industry and academia. The distinction between HW and SW is disappearing. What is in software in one product could be in hardware in the next version of the same product. Therefore, designers of systems must be equally knowledgeable in both: SW and HW.

Likewise, academia must start teaching courses in which the implementation of computing and communication concepts is presented in SW and HW in the same course. That requires the restructuring of many courses, particularly in computer science and computer engineering programs. In the long run, the computer science and computer engineering departments at our universities will be combined.

Similarly, the disappearing difference between SW and HW requires changes in the organizational structure of system design and manufacturing companies in any application domain. The present separation of application, system design, and SW and HW groups is not efficient. Likewise, the current simulation-based design flow is not sustainable since it requires designers to first learn the modeling language, develop models, and then verify the design, which all together takes too long and decreases design productivity. The system-design methodology of the near future must be based on solid scientific principles to enable its easy use and deployment.

In addition, the design and tool community must start changing their view of generic design practices to concentrate on applications' embedded systems and their specific needs. The 1000x productivity will not come without some serious experiments. After several years of work, we are glad to see the light at the end of the tunnel. In the succeeding chapters, we will present system design practices that have been proven in some real success stories though always with the understanding that more work lies ahead.

Chapter 2

SYSTEM DESIGN METHODOLOGIES

In this chapter we will look at different design methodologies, or design flows, for multi-processor systems. Design methodologies have evolved together with manufacturing technology, design complexity, and design automation. Improvements in technology have increased design complexity to the point that designers are no longer capable of making complex designs manually. To solve this problem, design automation tools, also known as computer-aided design (CAD) tools, were introduced. In order to make CAD tools more efficient and design algorithms more manageable, design-automation researchers as well as tool developers were forced to introduce more stringent design rules, parameterize components and minimize component libraries. As design complexities continued to increase, tool developers created new design abstraction levels and tried to use the same design strategy from the circuit level, to the logic level, to the processor level, and finally to the system level.

In this chapter, we will explain some basic system design methodologies related to the different abstraction levels in the Y-chart we introduced in Chapter 1.

2.1 BOTTOM-UP METHODOLOGY

Bottom-up methodology started even before CAD tools were invented. It is still in use in much of the industry today, at least partially, because it follows an intuitive methodology of building parts before assembling the whole product. In a typical bottom-up methodology, designers develop components and then store them in a library for use on the next-higher abstraction level.

As we can see in Figure 2.1, we have libraries of transistor, logic, RTL, and processor components. Components in each of the libraries are used to build

D.D. Gajski et al., *Embedded System Design: Modeling, Synthesis and Verification*,
DOI: 10.1007/978-1-4419-0504-8_2,

components in the library on the next abstraction level. On the Circuit level, we use transistors and develop circuits and their layouts for the basic logic components such as gates, flip-flops, bus drivers, and others. These components become standard cells for higher level design and layout tasks. These standard cells, with their functionality, structure, and layout, are stored in the Logic component library for use on the Logic level in Figure 2.1. On the logic level, we create register-transfer components such as registers, register files, ALUs, multipliers, and other components for processor micro architecture using Boolean expressions or FSM and FSMD models. After logic synthesis of these RTL components, we perform the placement and routing with standard cells for each component and store them in the RTL component library. On the Processor level, we start with C code or an Instruction set and generate the structure of processing elements (PEs) or communication elements (CEs). At this level we also perform floorplanning, placement, and routing of these PEs or CEs using the components from RTL library, and store them in Processor library. On the System level, we start with a model of computation (MoC) and generate the system structure consisting of multiple PEs and CEs from the Processor component library. Finally, we perform the system layout by using the component layouts from the Processor library. Note that each component library has functional, structural, and layout models for each component in the library. So by creating components and storing them in libraries, we can then apply them in each successive abstraction level.

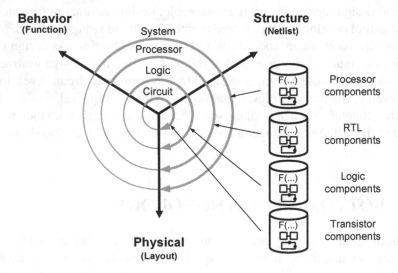

FIGURE 2.1   Bottom-up methodology

The advantage of bottom-up methodology is that abstraction levels are clearly separated, each with its own library. This allows for globally-distributed locations for design on each abstraction level, and for easier management of design on each abstraction level, since each group supplies a component library for the next level of abstraction. The disadvantage of this approach, however, is that the libraries must include all possible components with all possible parameters and that these must be optimized for the metrics required by any present and possible future applications. This is a very difficult and never-ending task since it is very difficult to anticipate on the lower abstraction level all the needs on the next higher abstraction level.

2.2 TOP-DOWN METHODOLOGY

In contrast to bottom-up methodology, top-down methodology does not attempt a component or system layout until the entire design is finished. A top-down methodology begins with a particular MoC and generates from it a system platform or system structure in which every component has its parameters and required metric values defined, but not its structure or layout. On the next level of abstraction, each PE or CE component is further decomposed into smaller RTL components. For example, in Figure 2.2, PE and CE components that were generated on the System level are decomposed into RTL components with their parameters and required metrics defined.

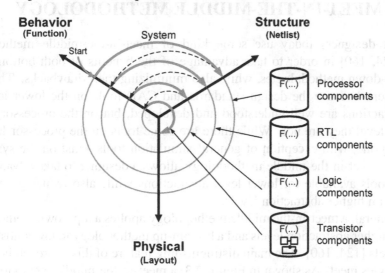

FIGURE 2.2 Top-down methodology

In this case, each functional unit, such as the ALU, has all its functions specified, as well as its delay and power requirements. After those are determined, each of the RTL components is further decomposed into logic components or gates. Finally, each logic component is broken down into a transistor netlist, in which each transistor layout represents a basic cell. All such basic cells, for the entire system, are placed on silicon and connected accordingly using placement and routing methods and tools. Such top-down methodologies were in use in design of early computers but today's designs are too complex for such a complete top-down methodology.

In general, top-down methodology leaves placement and routing for the last step by avoiding the layouts on other levels of abstraction. Unfortunately, the system and component metrics are not known until the last step and therefore it is very difficult to optimize the whole design. The design decomposition or synthesis has to be repeated over and over again without designers really knowing whether optimization is going in the right direction. In order to avoid too many design iterations, designers need the concept of metric closure in which different metric values from lower levels of abstraction are used to annotate design on higher level of abstraction. In this case designers can estimate optimized metric values on the lower level of abstraction during the next design iteration on the higher levels of abstractions. Unfortunately, metric closures are difficult to achieve since metric estimations are as difficult as performing real designs.

2.3 MEET-IN-THE-MIDDLE METHODOLOGY

Most designers today use some kind of meet-in-the-middle methodology [124, 160] in order to take advantage of the benefits of both bottom-up and top-down methodologies, while also minimizing their drawbacks. This is convenient because the design standards and CAD tools on the lower levels of abstractions are well understood and developed, but on the processor and system level they are not. While there are some tools on the processor level, almost none, with exception of general simulation tools, exist on the system level. A meet-in-the-middle methodology allows a designer to take advantage of the tools available for lower level abstractions while also reducing design layouts on higher abstraction levels.

In general, a meet-in-the middle methodology applies a top-down methodology to higher abstraction levels and a bottom-up methodology to lower abstraction levels [124, 160]. The main distinguishing feature of this approach is how these styles meet. As shown in Figure 2.3, a meet-in-the-middle methodology could start with a MoC and synthesize the system platform with virtual PEs and CEs which are after that synthesized with RTL components from the RTL library. These PEs and CEs also include commercially available IPs which are also supplied as netlists of RTL components. Therefore, all PEs and CEs are decomposed into RTL components from the library. Each RTL component has

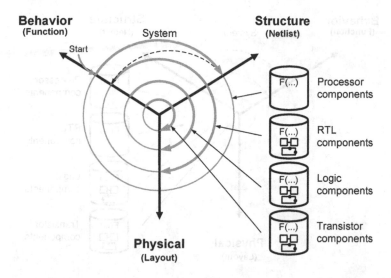

FIGURE 2.3 Meet-in-the-middle methodology (option 1)

its own structure and layout layout generated through some bottom-up methodology in the library. These RTL component layouts are combined through floorplanning, and routing into the layout of the multi-core platform. For example, ALU components in such a library would be limited to lengths of 8, 16, 32, 64 bits, and nothing in between.

Therefore, with such a meet-in-the-middle methodology, we do physical design or layout three times: once for standard cells, a second time for RTL components and a third time for the entire system platform, using the layout of these RTL components. This mixed methodology has the advantages of both bottom-up and top-down methodologies, since RTL components with their metrics are available from the libraries while the system is synthesized top-down from the RTL components. However, this approach has the drawback of requiring designers to do layout more than once. Moreover, system optimization is more difficult using already-made RTL components because they may not be tuned to the requirements of each PE or CE in the targeted system platform.

Another possibility for a meet-in-the-middle methodology would be to perform system layout with logic components or standard cells, as shown in Figure 2.4. As with the first meet-in-the-middle methodology we described, this one starts with a MoC and synthesizes the system platform with virtual PEs and CEs. Those PEs and CEs are then synthesized with RTL components, which themselves are further synthesized with logic components. Commercially available IPs that are described on the RTL level are also synthesized with RTL and logic synthesis tools that generate logic components netlists. Therefore, every IP component, as well as the synthesized PEs and CEs, are

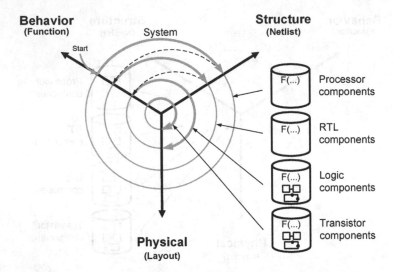

FIGURE 2.4 Meet-in-the-middle methodology (option 2)

decomposed into logic components from the Logic component library. Since each logic component has a layout as a standard cell, they are finally combined through floorplanning and routing into the layout of a multi-core platform.

In this case, we do physical design or layout only twice: once for generating standard cells and a second time for the entire system platform, using the standard cells layouts. This mixed methodology has an advantage in that only the standard cell layout has to be upgraded with the introduction of a new fabrication technology. The RTL component layouts, which are much more complex and in higher numbers do not need to be upgraded. An additional benefit is that the whole design is flattened to standard cells and the layout is performed only once. However, a system layout using standard cells is more complex than it would be with RTL components, and the design metrics are less predictable and controllable since standard cells for each RTL component may not be all in one place. Furthermore, using such inaccurate metrics makes it difficult to perform any system optimization on higher abstraction levels.

2.4   PLATFORM METHODOLOGY

The three design methodologies presented in the previous sections represent ideal cases of three different design concepts. In reality, design methodologies differ from company to company and even between different groups in the same company. They are also very much product oriented [165]. In this case, system design usually starts with an already-defined platform, usually one defined by a

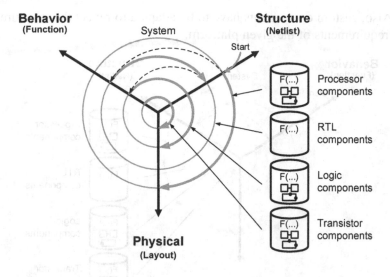

FIGURE 2.5 Platform methodology

well-known platform supplier or defined locally inside the company as shown in Figure 2.5. Such platforms may have already some standard components, such as memories and standard processors with well-defined layouts. The system platform may also be upgraded with the addition of custom components that will be synthesized with processor and RTL synthesis tools, after which the layout of these custom components can be obtained through standard cells. Furthermore, imported IPs are also converted to standard cell layout. Therefore, every custom component or imported IP can be defined with a netlist of standard cells, which is combined with netlists of other custom components for the combined standard cells layout. Such standard cell layout is then combined on the System level with layouts of standard processor and memory components into the system platform layout. When using such a platform, we perform physical design or layout three times: once for standard cells, then we use standard cells for the layout of custom components, and finally we use processor component layouts for the final platform layout.

In order to simplify platform design, some platforms have system layout for all standard processor components finalized with some space left open for the standard cell layout of custom components. When using such a platform, therefore, we perform layout only two times: once for standard cells and second time we use standard cells for the layout of custom components.

This mixed methodology has advantages from both bottom-up and top-down methodologies since standard processor components are available from the libraries and custom components can be inserted for application optimization. However, this approach has the weakness of requiring us to do layout more than once. Also, custom components have to be adapted to reflect the structure and layout requirements of the given platform.

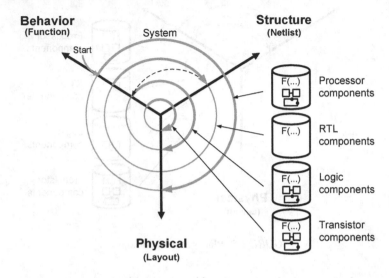

FIGURE 2.6 System methodology

The platform methodology can be upgraded to a system-level methodology by introduction of standard architecture cells and retargetable compilers. An architecture cell contains a parametrizable programmable processor such as one shown in Figure 1.5. The parameters include number, type and size of components, component connectivity, and the number of pipeline stages in the functional units, controller and the datapath. Such a standard architecture cells can be pre-synthesized with standard cells and inserted into the library of Processor components or generated on demand. A typical system-level methodology based on such architecture cells is shown in Figure 2.6. It starts with a MoC and generates the platform architecture consisting of standard or custom architecture cells. Since all the architecture cells have the layout model in the library the final system layout is obtained by combining the layouts of architecture cells.

This methodology has advantage of dealing only with two highest abstraction layers. Therefore, it is well-suited for application experts with minimal knowledge of system and processor design. However, it requires a retargetable compiler to cover different architecture cells.

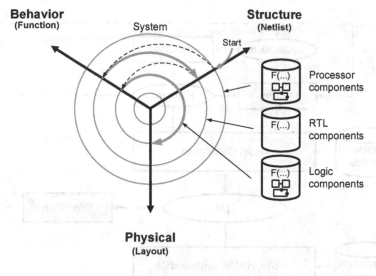

FIGURE 2.7 FPGA methodology

2.5 FPGA METHODOLOGY

Field-Programmable-Gate-Array (FPGA) methodology is based on the FPGA substrate, which consists of a multitude of 4-bit ROM cells called Look-up Tables (LUTs). These LUTs can implement any 4-variable Boolean function. Therefore, in this methodology, every RTL component in the RTL component library must be decomposed into these 4-variable functions. Then, the Processor components are synthesized out of available RTL components.

In other words, a FPGA methodology shown in Figure 2.7 uses a top-down methodology on both the System and Processor levels, in which standard and custom PEs and CEs are all expressed in terms of LUTs. A system design starts by mapping an application onto a given platform and then synthesizing custom components down to RTL components which are defined in terms of LUTs. Standard processors components in the Processor library are already defined in terms of LUTs. Once all components in the platform are defined, we flatten the whole design to LUTs and BRAMs and perform the placement and routing with the tools provided by FPGA suppliers.

This type of top-down system design has the same weaknesses as any top down methodology in that it is difficult to optimize the whole design by flattening the whole design just to basic LUT cells. Furthermore, designers do not know how the FPGA supplier-provided layout tools will map and connect all the LUTs and BRAMs.

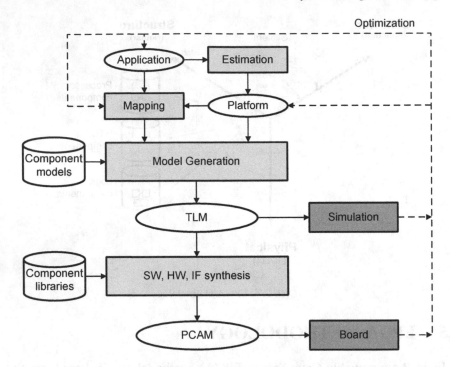

FIGURE 2.8 System-level synthesis

2.6   SYSTEM-LEVEL SYNTHESIS

In the previous sections we described several basic strategies in system design. However, system design flow has been changing alongside fabrication technologies and automation tools over the last 50 years. The changes started with lower levels of abstraction, which are well understood today. However, the higher levels of abstraction are still under investigation and discussion. In this section and the next, we will describe briefly the synthesis process from a behavioral description to a structural description on system and processor levels.

As shown in Figure 2.8, system-level synthesis starts with an application written in some MoC such as a set of sequential and parallel processes communicating through message-passing channels. Such a MoC must execute on a platform of multiple standard and custom processors connected through an arbitrary network. This type of platform can be defined partially or completely after estimating some characteristics of the application in terms of performance, cost, power, utilization, configurability, and other considerations. Platform definition can be done manually or automatically.

Once the platform is defined, an application must be partitioned and each partition assigned to a processor or IP in the platform. In order to verify that the application executes on the platform and satisfies all the requirements, we need to generate a simulatable and possibly verifiable model such as a timed Transaction-Level Model (TLM). After simulation, the design can be optimized if it does not satisfy the requirements by changing the platform, the application code, or the algorithms used in that code. We can also change the mapping of the application to the platform. For example, we can minimize external communication by grouping heavily communicating processes and assigning the whole group to one processor. It is also possible to assign performance-demanding processes to different processors or specialized IPs, or to pipeline performance-demanding processes if possible.

After we obtain a satisfactory application code, platform, and mapping, we can synthesize each component. Three types of components are needed: custom SW, HW, or IF components. SW components are for scheduling of processes such as different types of RTOS, and for communication and interfacing across the platform. HW components are various custom processors and custom hardware units, as described in the previous chapter. We also need communication components such as bridges and transducers for protocol conversion, and interface components such as bus arbiters and interrupt controllers.

Having synthesized these platform components, we need to generate a CAM model that contains binaries for downloading to processors and RTL descriptions for the HW parts in the platform. This can be done automatically or manually. Such a CAM is downloadable to standard FPGA boards for system prototyping, whose results can be used for final optimization of the whole design.

Details on each of these tasks will be given in the chapters that follow.

2.7 PROCESSOR SYNTHESIS

On the processor level, the components are synthesized as standard processors, custom processors, and custom hardware units, which are sometimes called IPs. The standard and custom processors are usually defined by their instruction sets. Custom processors can be also defined by the algorithm or the programming language code that they execute. They are programmable so that new algorithms and the code can be added or existing one modified. Custom hardware units or IP are usually not programmable. They are used as accelerators to execute special functions for a particular application, such as multimedia applications.

As shown in Figure 2.9 the synthesis process starts with a given Specification in a programming language, which is compiled into some Tool model such as

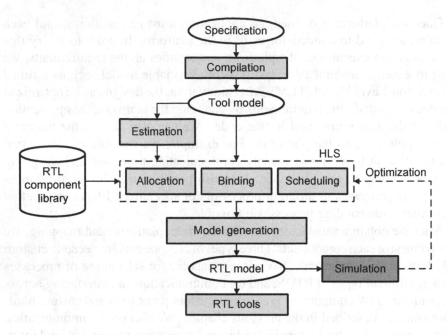

FIGURE 2.9 Processor synthesis

CDFG or a FSMD or a three-address code. This formal model can be used for Estimation of the future processor architecture and its metrics. It can be also used for some partial or complete allocation, binding, and scheduling. Processor synthesis, sometimes called High-Level Synthesis (HLS), takes the formal model and performs Allocation, Binding and Scheduling. The Allocation task selects necessary and sufficient components from the RTL component library and defines their connectivity. The Binding task defines binding of variables to registers, register files, and memories, operations to specific functional units and register-to-register transfers to specific buses. Scheduling assigns operations and register transfers to clock cycles. These three tasks compete with each other, so a completely optimized design is not easy to achieve. That is why estimation and pre-HLS comes handy. Pre-allocation helps in partial or full definition of processor architecture. This way we can avoid the clock-cycle estimates; since many or all of the register-to-register delays are known ahead of time, there is no need to wait until the end of HLS to find out the clock cycle time. Pre-binding may bind frequently-used variables to fast registers, register files, or a scratch-pad memory to avoid lengthily delays caused by loading and storing to the main memory. Pre-scheduling can assign key inner loops to high-speed, pipelined functional units or it can pre-schedule such loops to specific paths in a pipelined datapath.

Once HLS is finished we need to generate a RTL Model of the processor that can be synthesized with standard RTL synthesis tools.

2.8 SUMMARY

In this chapter, we explained the differences between top-down, bottom-up, and meet-in-the-middle methodologies by exposing their taxonomizing structures. We also highlighted some features of key ASIC and FPGA methodologies. We also detailed one methodology branch, synthesis, explaining how the process might work on both the processor and system levels.

However, we have to acknowledge that there are many more design methodologies, almost one for every group, product, and company [103, 47, 63, 100, 129, 195, 184]. They may start with different specifications, may use other models for verification of different concepts and metrics, and they may need a different type of outputs. However, all design methodologies must address the basic system needs and issues we have introduced in this chapter These methodology issues will be discussed in more detail in the succeeding chapters.

Chapter 3
MODELING

At the core of any design methodology are models at various steps of the flow. Models provide an abstract view of the design at any given time, representing certain aspects of reality while hiding others that are not relevant or not yet known. As such, design models at each level of abstraction provide the basis for applying analysis, synthesis or verification techniques. However, as discussed in previous chapters, tools for automation of these design processes beyond simulation can only be applied if models and corresponding abstraction levels are well-defined with clear and unambiguous semantics. In doing so, modeling concepts and techniques can have a large influence on the quality, accuracy and rapidity of results. Hence, modeling is concerned with defining the level and organization of detail to be represented, i.e., the objects, composition rules and, eventually, transformations, such that meaningful observations can be made, desired requirements are explicitly specified, and design automation tools can be applied.

As explained in previous chapters, system behavior is generally described as a set of concurrent, hierarchical processes that operate on and exchange data via variables and channels. On the other hand, a system platform consists of a set of system components connected by a network of busses. Components in a platform realize system behavior by executing processes, storing variables and sending messages over channels. Hence, various components provide different aspects of system computation and communication, whether in software, hardware, or in a combination of both. In system modeling, we therefore need to develop models at varying levels of detail. Furthermore, we have to define models for each component as well as for the whole system.

As discussed previously, different models are needed at different steps in the design process. There are abstract system models for application designers who must develop algorithms and verify that they will work correctly on the

D.D. Gajski et al., *Embedded System Design: Modeling, Synthesis and Verification,*   49
DOI: 10.1007/978-1-4419-0504-8_3,
© Springer Science + Business Media, LLC 2009

platform under the given constraints. More detailed models are needed for system designers who must architect the platform The most detailed model is needed for implementation designers who need to verify correctness of the software and hardware implementation.

In this chapter, we will discuss concepts and techniques for modeling of systems at various levels of abstraction. We first present Models of Computation (MoCs) and design languages, which together provide the foundation for defining system behavior and models throughout the design flow. Based on these general principles, we then describe details of computation and communication modeling in the system components. The basic system component for computation is a processor with functionality that can be separated into application, operating system, hardware abstraction and hardware layers. Communication functionality can be modeled as stacks of network and protocol layers that are inserted into processors and CEs to realize drivers and interface hardware. In the end, we will show how these concepts and layers are combined into system models for application, system, and implementation designers in the form of a Specification Model (SM), a Transaction-Level Model (TLM) and a Cycle-Accurate Model (CAM), respectively.

3.1 MODELS OF COMPUTATION

A Model of Computation (MoC) is a generalized way of describing system behavior in an abstract, conceptual form [130, 101, 117]. As a result, MoCs are the basis for both humans and automated tools to reason about behavior and the requirements and constraints of computations to be performed. Typically, MoCs are represented in a formal manner, using, for example, mathematical functions over domains, set-theoretical notations, or combinations thereof. This establishes a well-defined semantics and allows formal techniques to be applied. Different MoCs can thereby have various degrees of supported features, complexity and expressive power. Hence, the analyzability and expressiveness of behavioral models is in the end determined by their underlying MoCs.

MoCs are generally based on a decomposition of behavior into pieces and their relationships in the form of well-defined objects and composition rules. In the process, MoCs are inherently tied to abstracted definitions of functionality, i.e., processing of data, and order, i.e., notions of time and concurrency. Models of time at higher levels of abstraction typically define a partial order in which a relative sequence of concurrent executions is only specified for a subset of the events in the system, purely based on causality and inherent dependencies. In a physical implementation at lower levels, by contrast, every event is attached to a precise instant in real time, which imposes a total order on the execution of the system. To define the order, composition rules establish the dependencies between objects in the form of data and/or control flow. Examples at either end of the spectrum can include shared variables for unordered data flow or synchronization mechanisms such as events for data-less control flow and ordering only.

Arguably the most common MoC is an imperative model, as realized by sequential programming languages such as C or C++. In an imperative MoC, behavior is described as a sequence of statements that operate on and change program state. Imperative models can be graphically represented in the form of flow charts or activity diagrams [23]. Both statements and state can be decomposed into hierarchical structures using procedural or object-oriented programming methods. Statements communicate solely through manipulations of shared memory. For that reason, statements are strictly ordered in time based on the sequence in which they are defined. Note that modern compilation techniques can relax this requirement and extract concurrency or optimize the state space by splitting imperative code into basic blocks and abstracting inter- and intra-block dependencies into CDFGs. In contrast, functional or logical programming models follow a declarative style and are directly based on variants of a dataflow MoC (see Section 3.1.1 below) with ordering based on explicit dependencies only.

All imperative, functional or logical programming models describe the transformative aspects of systems as pure functionality that maps inputs to outputs. In embedded systems, however, time is usually a first-order property. Such systems are reactive in the sense that they continuously interact with their environment. The relationship, relative ordering and interleaving of outputs and inputs is part of the definition of their behavior. Therefore, so-called synchronous languages [18] follow an approach where concurrency and ordering is explicitly specified in the code instead of relying on extracted or implicit scheduling of operations. Program statements are composed into concurrent blocks that communicate through signals to exchange sequences of values and events. Furthermore, such languages divide the time model into a sequence of discrete steps and mandate that all operations and events within each step happen simultaneously and instantaneously, i.e., in zero time at the ticks of a set of logical clocks. A conceptually discrete time model where all delays assumed to be zero establishes a total order and makes synchronous languages fully deterministic, allowing for proofs of correctness. Examples of synchronous languages include Esterel [21], which follows an imperative style to define block behavior and is based on an underlying finite state machine MoC (see Section 3.1.2). By contrast, Lustre [85] follows a functional (declarative) style based on a dataflow model (see Section 3.1.1) in which all blocks execute concurrently and in lockstep.

On top of basic, fine-grain programming models that are composed out of objects at the level of individual statements or operations, higher-level MoCs can be defined to reason about interactions between complete coarse-grain blocks of code. Such MoCs can be broadly subdivided into process-based and state-based models. Process-based models are data oriented and are typically used in system behavioral models to describe desired application functionality. State-based models, on the other hand, focus on explicitly exposing and representing control flow. They are used for control-dominated applications and for modeling of designs at the implementation level. Throughout the design flow, a variety of such MoCs can then be used to describe designs. Note, however, that MoCs only capture behavioral aspects. Any system model will therefore have to combine MoCs with capabilities to represent structural aspects of the design as well.

3.1.1 PROCESS-BASED MODELS

Process-based MoCs represent computation as a set of concurrent processes. Processes are internally described in an imperative form using sequential programming models. In other words, the overall system is modeled as a set of blocks of code that execute in parallel and are generally independent of each other. Thus, process-based MoCs focus on explicitly exposing available concurrency. They are untimed and ordering is only limited by data flow between processes as is the case, for example, in a producer-consumer type of relationship. As such, they are applicable for modeling of functionality at the input of system design flows, specifically for streaming applications where interactions are dominated by data dependencies.

Different process-based MoCs then vary in the semantics of communication they support to exchange data and establish dependencies between processes. As realized by various operating systems (e.g., Posix threads [29]), languages (e.g., Java threads [79]) or parallel programming environments (e.g., the Message Passing Interface, MPI [80]), general-purpose process models typically support a broad set of Inter-Process Communication (IPC) mechanisms with universal semantics. Low-level and implementation-oriented thread-based models are built on shared memory and shared variable semantics with the subsequent need for additional mechanisms (such as semaphores, mutexes or critical sections) to explicitly synchronize accesses to shared resources [119]. Alternatively, in message-passing models, each process has a separate local memory space and processes exchange blocks of data in a synchronous, rendezvous-style or asynchronous, queue-based fashion. In the synchronous case, message senders are always blocked until the receiver is ready to accept the data. In the asynchronous case, messages are buffered and senders may or may not block, depending on the buffer fill state.

Definitions of concurrency and communication in process-based models directly translate into properties such as deadlocks and determinism. Deadlocks can arise if there is a circular dependency between two or more processes where each process holds an exclusive resource that the next one in the chain is waiting for. For example, a process might wait for a semaphore that is blocked by another process and vice versa. Deadlocks can be prevented or avoided by statically ensuring that chains can never occur or by dynamically breaking them at runtime.

Determinism is related to the outputs of a model for a given set of inputs. If a model is deterministic, the same inputs will always produce the same results. By contrast, if a model is non-deterministic, its behavior is, for at least some inputs, undefined. Note that non-determinism is different from random behavior. In the random case, different outputs will appear with a certain probability, whereas non-determinism will not give any guarantees at all. Non-determinism makes it hard to ensure that the behavior is correct if a specific result is desired. Especially during validation, it is generally not feasible to produce all possible outcomes. Randomized simulations can alleviate this problem yet still not provide guarantees. A fully deterministic model, on the other hand, will guarantee results but might instead lead to overspecification. For example, truly concurrent processes have to be non-deterministic in the order in which they execute. This provides an implementation with the necessary degree of freedom to choose a specific schedule.

To cope with these issues and propose varying solutions, different process-based MoCs have been developed over the years. Depending on their rigor, the most common process-based models can be roughly subdivided into process networks, dataflow models and process calculi.

PROCESS NETWORKS

Specialized process-based MoCs have been proposed that provide deterministic properties on a global scale while still allowing for non-deterministic execution of individual processes. This is generally achieved by ensuring that the order of process execution cannot affect overall behavior of the system. For example, in a Kahn Process Network (KPN) [104], processes are only allowed to communicate via uni-directional and point-to-point asynchronous message-passing channels, where messages (also called tokens) can be of arbitrary type. Channels are unbounded, and as such, senders can never block. Conversely, receivers always block until a complete message is available. Since processes can only wait for a single channel and cannot check whether data is available without blocking, they have to decide in each step whether to wait for a channel and which channel to wait for next. Therefore, the sequence of channel accesses is predetermined and processes cannot change their behavior depending on the order in which data arrives on their inputs. Hence, the behavior of the overall system is deterministic and does not depend on the order in which processes are scheduled. Note that a KPN can have deadlocks but is defined to regularly terminate on a global one when all processes are blocked while waiting for messages. Again, global deadlock and termination conditions do not depend on the chosen schedule.

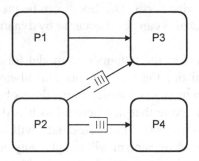

FIGURE 3.1 Kahn Process Network (KPN) example

In a KPN, however, the chosen scheduling strategy will influence other properties such as completeness or memory requirements. For example, Figure 3.1 shows a simple KPN with two processes *P1* and *P2* producing data that is consumed by process *P3*. In addition, a fourth process *P4* only depends on data from *P2*. In such a pure KPN model, processes are connected via unbounded FIFO queues with infinite buffers. Any KPN implementation, on the other hand, must run within the limited physical memory of a real machine. In this respect, the order in which processes are executed will determine the amount of memory needed. For example, if processes are executed in a round-robin fashion but *P1* and *P2* produce tokens at a faster rate than they can be processed by *P3* and *P4*, tokens will unnecessarily accumulate on the arcs. This can be avoided by only running processes whenever their data is needed. In such demand-driven scheduling, arcs between processes are essentially treated as synchronous message-passing channels with zero buffering. Note that this can create unnecessary backwards dependencies, which can potentially lead to artificial deadlocks.

Consider a variant in which *P3* does not consume any tokens at all or is blocked in a local deadlock with another process. In this case, a demand-driven scheduling would not execute *P2*, effectively blocking *P4* and the independent stream in between them as well. In contrast, a data-driven scheduling runs processes whenever they are ready. It would keep *P1*, *P2* and *P4* running but would also indefinitely accumulate tokens on the *P1-P3* and *P2-P3* arcs.

In general, KPNs are Turing complete and it is undecidable by any finite time algorithm whether they terminate (halting problem) or can at least run in bounded memory. Not being able to determine if and in what order processes have to be run when reaching full write buffers, any scheduling strategy must choose between a complete or a bounded execution. A complete execution runs processes as long as they are ready but might require unbounded memory. A bounded execution imposes limits on buffer sizes and will block senders when reaching buffer limits. Thus, a bounded execution may be incomplete and may potentially create artificial deadlocks leading to early termination. A data-driven scheduling algorithm prefers completeness and hence non-termination over boundedness. A demand-driven scheduling prioritizes boundedness over completeness and even non-termination. In practice, hybrid approaches are employed [154]. In Parks' algorithm, processes are executed until buffers become full, gradually increasing buffer sizes whenever an artificial global deadlock occurs. As such, the algorithm prefers non-terminating over bounded and bounded over complete execution. Note, however, that there are KPNs where a complete, bounded schedule exists that neither algorithm will find [68].

DATAFLOW

Overall, KPNs generally require both dynamic scheduling with runtime context switching and dynamic memory allocation. For these reasons, their practical and efficient realization is difficult to achieve. To improve on the short-comings of KPNs, extensions with restricted semantics have been developed. In a dataflow model, processes are broken down into atomic blocks of execution, called actors. Avoiding the need for context switches in the middle of processes, actors execute, or fire, once all their inputs are available. On every execution, an actor consumes the required number of tokens on all of its inputs and produces resulting tokens on all of its outputs. In the same way as KPNs, actors are connected into a network using unbounded, uni-directional FIFOs with tokens of arbitrary type. More formally, a dataflow network is a directed graph where nodes are actors and edges are infinite queues. Dataflow networks are deterministic and have the same termination semantics as KPNs.

Dataflow models map well onto concepts of block diagrams with continuous streaming of data from inputs to outputs. As a result, they are widely used in the signal processing domain and as the basis for many commercial tools such as LabView [96] and Simulink [95]. However, in their general form, questions as to their schedulability, boundedness and completeness, remain. For example, termination due to deadlocks is typically not desired in these types of applications, yet non-termination cannot be analyzed or guaranteed. Therefore, variants of dataflow that further restrict semantics of atomic execution have been developed. For example, Synchronous Data Flow (SDF) [120] models have found widespread adoption. In an SDF graph, the number of tokens consumed and produced by an actor per firing is constant and fixed. Hence, the amount of data flowing through the system is predetermined and can not dynamically change depending on, for example, elapsed time or received token values. Therefore, the graph can be statically scheduled in a fixed order. As a consequence, statically scheduled SDF graphs are bounded and required buffer

sizes are known before runtime. Note, however, that the choice of schedule
might still influence overall memory requirements.

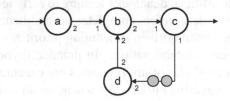

FIGURE 3.2 Synchronous Data Flow (SDF) example

Figure 3.2 shows an example of a simple SDF system with four actors, a, b,
c and d. On every execution, actor a produces two tokens; actor b consumes
three tokens (one on the arc from a and two on the arc from d) and produces
two (on the arc for c); actor c consumes one of b's tokens and sends a token to
d; and finally, actor d both consumes and produces two tokens on each of its
input and output arcs. Note that the graph is initialized by placing two tokens
on the arc between c and d. Such initialization tokens are necessary to resolve
any deadlocks that might exist in the raw graph, as is the case for this example.

To schedule such an SDF graph, we first determine the relative execution
rates of actors by solving the system of linear equations relating production and
consumption rates on each arc. For the example shown in Figure 3.2, we get
so-called balance equations

$$2a = b$$
$$2b = c$$
$$b = d$$
$$2d = c$$

which reduce to $4a = 2d = c = 2b$. Picking the solution with the smallest
rates, we have to execute c four times and b and d each two times for every
execution of a. Note that if the system of linear equations is inconsistent and
not solvable other than by setting all rates to zero, the SDF graph can not be
statically scheduled and would otherwise (if scheduled dynamically) lead to
accumulation of tokens.

After computation of execution rates, we can determine a schedule to be
executed periodically by simulating one iteration of the graph until its initial
state is reached again. Note that if a deadlock is reached during this process,
initialization tokens (as described above) will have to be placed on some arcs
for a valid schedule to exist. Any scheduling and simulation algorithm can then
be used to determine a firing order and many different schedules can usually be
generated for each graph. Schedules will vary in the sizes of buffers required
during their execution.

For example, a simple list scheduling of Figure 3.2 could result in an actor order of *adbccdbcc*. This schedule will accumulate at any time a maximum of 2 tokens on each arc for a total memory requirement of 8 token buffers. By contrast, a schedule of *a(2db)(4c)* would require a total of 12 token buffers, but would potentially result in a smaller code size as the actors *db* and *c* are executed within local loops. In general, depending on code generation and compiler optimizations, a single-appearance schedule in which each actor invocation appears only once (such as the second one above) can lead to a reduction in code size, potentially at the expense of buffer requirements. All in all, SDF approaches allow for efficient implementation of models in which dependencies between blocks can be statically fixed.

PROCESS CALCULI

A further restriction of dataflow models and a formalization of process-based execution into a sound mathematical calculus framework is provided by models such as Communicating Sequential Processes (CSP) [90] or the Calculus of Communicating Systems (CCS) [141]. As in a demand-driven scheduling of KPNs, communication between processes in such models is limited to rendezvous-style, synchronous message-passing. Similar to the concept of a model algebra introduced in Chapter 1, this strict semantics allows an algebra of processes to be developed based on a definition of corresponding objects, operations and axioms. Objects in a process algebra are processes $\{P,Q,...\}$ and channels $\{a,b,...\}$. Operations are process compositions such as parallel $(P \parallel Q)$, prefix/sequential $(a \rightarrow P)$, or choice $(P + Q)$ operators. Finally, axioms define basic truths such as indemnity $(\oslash \parallel P = P)$ or commutativity $(P + Q = Q + P \text{ or } P \parallel Q = Q \parallel P)$. Models can then be written as process algebraic expressions and manipulated or compared, e.g., to prove equivalence, by successively applying axioms or derived theorems. Due to their rigorous semantics, process calculi have been used as the basis for many parallel programming or design languages, among them OCCAM [121] or Handel-C [38], both of which are based on a CSP model.

In summary, process-based models have the general advantage of explicitly exposing concurrency by focusing only on dependencies due to the flow of data through the system. Implementation-oriented solutions have the fewest restrictions but also provide little to no guarantees or opportunities for analysis and optimization. At the other end, SDF models are statically fixed and can be implemented very efficiently. Note that Data Flow Graphs (DFGs), as used, for example, to represent dependencies in expressions or basic blocks of a CDFG, are further restricted variants of SDF in which graphs have to be both directed and acyclic, and actors representing operations are only allowed to produce and consume a single value per arc and firing. On the other hand, extended variants of SDF, such as Boolean [118] or Cyclo-Static Dataflow [22], relax some of the restrictions in order to become Turing complete or increase the scope. In between implementation-oriented and static models, KPNs and, to a more limited extend, process calculi provide even greater flexibility while still being at least partially analyzable (e.g., in terms of determinism) when modeling dynamic behavior, as found, for example, in many modern multimedia applications.

3.1.2   STATE-BASED MODELS

State-based models generally describe behavior in the form of states and transitions between states. As such, they are primarily focused on an explicit representation of the status of computation at any time, where a state is a snapshot of the union of all memory and essentially reflects history. In addition, state-based models explicitly represent the flow of control as transitions between different states Imperative models, flow charts and CDFGs, by contrast, only encode state implicitly (in the form of associated global variables).

State-based models were originally developed to describe the stepwise operation of a machine in an abstracted, formalized manner. Specifically, they are almost exclusively the basis for modeling of synchronous hardware down to a cycle-by-cycle level. In addition, state-based models are often used to specify the abstract behavior of control-dominated, reactive applications that are driven by actions in response to events.

State-based models that represent computation are usually finite in the number of states and transitions. As a result, they are not Turing complete. Yet, they are powerful enough to describe large classes of computation. In addition, their finite nature makes them amendable to analysis and optimization through formal methods to check, for example, equivalence, minimization or reachability of states.

FINITE STATE MACHINES

The most fundamental model of computer science is a Finite State Machine (FSM) or finite automaton. An FSM is formally defined as a quintuple

$$< S, I, O, f, h >$$

where S represents a set of states, I represents a set of inputs, O represents a set of outputs, and f and h are the next-state and output functions, respectively [62]. The next state function $f : S \times I \rightarrow S$ defines for every state and every input the transition to the next state of the FSM. An FSM is deterministic if there is one and only one next state for every input and state. On the other hand, an FSM is non-deterministic when f is a multivalued function. The output function *h* defines the output values of the FSM depending on the state and optionally input values. In a so-called Mealy FSM, the output function $h : S \times I \to O$ is transition-based and outputs are defined for every state and every input. In contrast, a Moore FSM is state-based and the output function $h : S \to O$ does not depend on the inputs but only on the current state. Note that a Moore FSM is equivalent to a Mealy FSM in which incoming transitions for every state have the same output. Therefore, a Mealy FSM can be converted into a Moore FSM by splitting states depending on the different outputs generated when entering the state.

An FSM can be efficiently stored in tabular form. However, FSM models quickly become too large to be processed by humans or tools and are useful for computations represented by several hundred states. On the software side, FSMs are often used as automata to recognize or represent language grammars or regular expressions. On the hardware side, FSMs are used as abstracted representations for analysis and optimization of sequential circuits that are implemented in the form of a state register, next state and output logic. In this case, FSM models are cycle-accurate and each state corresponds to one clock cycle.

The original FSM model uses binary variables for inputs and outputs, where function *h* assigns constants of 0s and 1s to output variables. Consequently, an FSM has to include a new state for every distinct condition to be encountered and remembered, e.g., when counting the number of times an event has occurred. To avoid this state explosion and reduce complexity, a Finite State Machine with Data (FSMD) introduces standard integer or floating point variables, which allows each state or transition to be associated with an expression over these variables [38]. Formally, a FSMD is a sixtuple

$$< S, I, O, V, f, h >$$

that extends the FSM definition with a set of variables *V* and modifies the next state function $f : S \times I \times V \to S \times V$ and output function $h : S \times I \times V \to O$ to define mappings that include variable values. Note that FSMs are a subset of FSMDs, i.e., every FSM is also a FSMD. Conversely, FSMDs can be translated into equivalent FSMs by expanding every possible variable value into a separate state.

Figure 3.3 shows an example of a counter modeled as a FSMD that increments a variable *v* whenever input event *c* occurs. The FSMD has three states *s1*, *s2*, and *s3* and seven transitions representing state changes under different inputs and conditions. In this case, start state *s1* initializes *v* to zero and then enters the waiting state *s2*. In state *s2*, the FSMD does not perform any operation, and it will stay in this state as long as *c* is zero. Once *c* becomes true, the FSMD transitions to state *s3* and continuously increments *v* until *c* goes back to zero, at which point the FSMD transitions back to *s2*. Finally, when receiving input

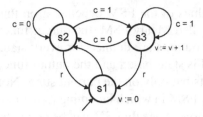

FIGURE 3.3 Finite State Machine with Data (FSMD) example

event *r* in either *s2* or *s3*, the FSMD is reset and restarted by transitioning back to state *s1*. As is the case for most embedded, reactive systems, the FSMD executes indefinitely and does not terminate. In general, a state machine can declare an explicit end state if it is meant to be embedded in a larger context.

The FSMD model is widely used to represent hardware implementations of RTL processors consisting of a controller and a datapath [3]. In this case, each state executes in one clock cycle. States and transitions of the core FSM thereby describe the implementation of the controller. On the other hand, variables, expressions and conditions describe the operations performed by the datapath in each cycle. In a similar manner, FSMDs can be used to provide a state-oriented view of imperative programming models. Transitions and states describe the control flow of the program where each state computes a set of expressions corresponding to the statements in the code. Note that in this case, the FSMD are usually not cycle accurate since states can represent whole basic blocks that may require several clock cycles to execute. Furthermore, note that imperative models are more vividly represented by a CDFG describing control and data dependencies between and within basic blocks, respectively.

HIERARCHICAL AND CONCURRENT STATE MACHINES

Hierarchy and concurrency are further mechanisms to manage complexity of the state space. In a hierarchical state machine, states can be complex, so-called super states, which internally consist of a complete state machine each. Consequently, individual FSMDs are hierarchically composed into a so-called Super State FSMD (SFSMD). In an SFSMD, entering a super state is equivalent to entering the start state of the SFSMD contained within. Super states can be exited by defining an end state in the child SFSMD. Whenever a super state reaches it end state and exits, the parent SFSMD will transition to and enter a specified other of its super states. As an alternative to explicit end states in children, a parent SFSMD can declare a transition between super states that will exit a child SFSMD whenever a specified condition becomes true, independent of which substate the child is in at that time. As such, hierarchy allows both to organize complexity and potentially reduce the number of transitions in the state diagrams.

Concurrency allows complex state machines to be decomposed into multiple, separate FSMDs running in parallel. Concurrent FSMDs can thereby communicate through a set of shared signals, variables and events. Interactions between state machines are usually based on a model that operates concurrent, communicating FSMDs in a synchronous, lock-step fashion. By ensuring that FSMDs all transition and update or check signals at the same time, it can be guaranteed that they will not miss each other's events and hence can safely exchange information.

When combining both hierarchy and concurrency, so-called Hierarchical and Concurrent Finite State Machine (HCFSM) models emerge, such as the ones pioneered in Harel's graphical StateCharts language [86] and used for Unified Modeling Language (UML) state diagrams [23]. In the original StateCharts, each hierarchical super state can be either a so-called AND- or OR-composition of substates. OR states are used to describe regular hierarchy in which a parent state is at any given time in either one (but only one) of its substates. In contrast, AND states describe a concurrent composition where being in a parent state means that the system is at the same time in all of its substates.

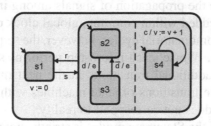

FIGURE 3.4 Hierarchical, Concurrent Finite State Machine (HCFSM) example

Figure 3.4 shows an example of a HCFSM as a variation on the counter FSMD presented earlier in Figure 3.3. At the top level, the system is modeled as an OR composition that starts execution in initialization state *s1*. Upon receiving the start signal *s*, the HCFSM enters the concurrent composition of state machines *s2*, *s3* and *s4*. The left state machine starts in state *s2* and essentially implements an edge detection that transitions between *s2* and *s3* and issues an event *e* depending on the presence or absence of event *d*. In parallel, *s4* implements a simple counter that increments *v* on every occurrence of *c*. At the same time, the hierarchical combination of *s2*, *s3* and *s4* can be aborted by an event *r* that transitions from whatever state the combined super state is in back to the start state *s1*.

As discussed above, HCFMs execute concurrent state machines in a lock-step, synchronized fashion. Different HCFSM models can vary in the details of their semantics, specifically depending on when and for how long generated events take effect. Introduced as a purely graphical notation, the original StateCharts description did leave many of these issues open. As a result, a wide variety of interpretations have been proposed over the years. Notably, the semi-official semantics as realized by Harel's own Statemate tool set follows an approach in which events that are posted in one step are valid in and only in the next step [87]. Together with additional rules about, among others, priorities of conflicting transitions, this makes Statemate models deterministic in their externally observable behavior.

Statemate thereby offers two different, so-called synchronous and asynchronous execution modes. In the synchronous mode, steps are executed at regular intervals, sampling all inputs, executing transitions and posting events for the next interval in each. This corresponds directly to a hardware implementation with a network of synchronous state machines connected in a Moore fashion. In the asynchronous mode, global steps can each consist of a sequence of microsteps, where microsteps are assumed to execute in zero time. External inputs are only sampled at the beginning of each global cycle while internal events are propagated through a chain of microsteps until the system stabilizes and no more events are generated. As long as there are no cyclic dependencies, this mode can emulate the propagation of signals among immediately reactive Mealy machines embedded within common, global clock cycles.

In the presence of combinatorial cycles, however, the sequence of microsteps might never terminate. Worse, each microstep performs state updates, which in turn might enable additional transitions in the next microstep, leading to superfluous or multiple transitions of state machines within each global step. Clearly, this behavior does not correspond to reality.

In all cases, none of the Statemate modes is strictly synchronous as required for precise modeling of interconnected Mealy machines. As discussed previously in the context of synchronous languages (at the beginning of Section 3.1), for models to be fully deterministic, events within each global cycle all have to occur at the same instant and within zero time [18]. In that case, combinatorial cycles can lead to global inconsistencies or non-determinism. To deal with such issues, truly synchronous languages either generally reject models with cycles at compile time (e.g., as is the case for Lustre [85]) or require that a unique fixed-point solution exists in every global step (as realized, for example, in Esterel [21]). Likewise, note that strictly synchronous variants of HCFSM models have been developed by providing a dedicated synchronous interpretation (Argos [125]) or by using the StateCharts-like notation as a graphical frontend for Esterel (SyncCharts [6]).

PROCESS STATE MACHINES

To avoid the need to maintain a global time, models exist that compose concurrent, communicating FSMDs asynchronously in the same manner as is done in process networks (see Section 3.1.1). This then requires more complex handshaking protocols or mechanisms such as message-passing in order for FSMDs to be able to communicate reliably. For example, while leaving many semantic details undefined, UML state diagrams [23] as yet another variant of HCFSMs are generally based on an unrestricted asynchronous execution model in which concurrent state machines must explicitly coordinate their execution through event queues wherever necessary, e.g., to synchronize accesses to shared variables. As such, UML state diagrams are in the general case non-deterministic. However, their state-based nature allows other formal models, such as Petri Nets [142], to be superimposed on such asynchronous HCFSMs. Similar to process calculi (see Section 3.1.1), state-oriented mathematical models like Petri Nets abstract away actual functionality and only focus on representing interactions and relationships necessary to analyze concurrency, synchronization, determinism and properties such as boundedness, reachability or liveness.

Combining synchronous and asynchronous approaches to concurrency, so-called Globally Asynchronous Locally Synchronous (GALS) models, such as Co-Design Finite State Machines (CFSMs) [11] have been proposed. GALS models maintain local clocks for FSMDs within each block yet allow different blocks in the overall system to progress independently. Such models match the typical clock distribution in modern, complex system architectures. Nevertheless, at the leaves of the hierarchy, behavior is still described in an implementation-oriented form as clocked state machines communicating over signals or wires.

Taking these ideas further, we can develop a sound combination of process- and state-based approaches by fully integrating concepts of process networks (see Section 3.1.1) into HCFSM models. For example, in a Program State Machine Model (PSM) [63], leaves of the hierarchy contain complete asynchronous processes described in a sequential, imperative programming language. In the original SpecSyn language, behavioral VHDL code was used as the basis for describing processes [185].

In a PSM model, such so-called program states can then be composed hierarchically following a HCFSM style. At each level, either a sequential state-machine or a concurrent but asynchronous composition of program states is supported. When entering a program state, execution either starts with the first statement of the process code or, in the case of a superstate, by entering the set of start states in the same manner as a HCFSM. In contrast to HCFSMs, however, processes and hence program states have an explicit and clean notion of completion. Superstates can be exited in two ways: a so-called Transition-

Immediately (TI) arc is equivalent to a transition in an HCFSM (such as *r* in
Figure 3.4) that originates from the superstate to one of its siblings and can
be taken at any time, independent of the internal sub-state(s) the superstate is
in. On the other hand, a Transition-On-Completion (TOC) arc is defined as a
transition to a sibling of the superstate that is taken at the same time that the
substates internally reach a declared end state. A leaf state completes whenever
its process exits (i.e., reaches its end or explicitly returns). Furthermore, the
end state of a concurrent superstate is reached when all of its substates have
completed.

As mentioned above, concurrent processes of a PSM run asynchronously
to each other at any level. In the original SpecSyn model, processes can only
communicate through a set of basic shared variables, events and signals. Thus,
models are generally non-deterministic and have to explicitly implement any
protocols necessary to synchronize and coordinate process execution. As a result, processes are generally a mix of computation and communication code. To
improve on this situation, later extensions of the PSM model support the separation of communication from computation into distinct objects. For example,
the SpecC model, while also being based on C instead of VHDL, introduced
the concepts of channels for encapsulation of communication and of clearly
defined interfaces as the boundary for separation of process from channel functionality [63].

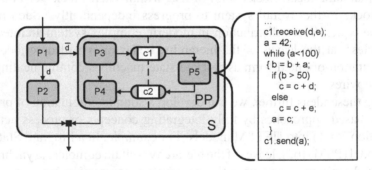

FIGURE 3.5 Process State Machine (PSM) example

Figure 3.5 shows an example of an extended Process State Machine (PSM)
model. At the leaves, the model consist of five processes, *P1* through *P5*, that
are described in standard C or C++ form. As before, the system *S* starts by
executing process *P1* and, depending on input *d*, either transitions to process
P2 or enters the concurrent superstate *PP*. Inside *PP*, the sequenc of process *P3*
followed by *P4* runs in parallel to process *P5*. Concurrent processes exchange
data by sending and receiving messages over channels *c1* and *c2*. *PP* completes
once both *P5* and *P4* are finished executing. Upon completion of either *P2* or
PP, *S* enters its end state, which transparently follows to TOC arc of *S* to one

of its siblings. Note that the example does not show any TI arcs, which were already previously seen in Figure 3.4. Nevertheless, in contrast to plain HCFSM models, clean completion semantics of PSMs results in a well-defined modular composability.

In summary, PSM models provide a powerful combination of both process-based and state-based concepts. Asynchronous process networks provide a means to describe dynamic, data-oriented application behavior limited only by the flow of data and data dependencies across computations. On the other hand, concepts of states and transitions allow explicit modeling of reactive, control-oriented systems in addition to providing a representation of implementation issues such as program state, data storage, operation scheduling or cycle-accurate behavior. At all levels, hierarchy and concurrency support organization and management of complexity through separation of concerns. In addition, separation of computation and communication supports further orthogonalization of concerns and enables coarse-grain, asynchronous concurrency and flexibility while still providing means, including libraries of message-passing or other communication channels, to maintain global determinism. All in all, combined PSM-type models are able to support the complete system design process all the way from specification of abstract system behavior down to cycle-accurate implementation of hardware or software components.

3.2 SYSTEM DESIGN LANGUAGES

In order for a design to be simulated, analyzed and verified by the designer, it needs to be represented in a formalized, machine-readable manner - that is, in some form of design language. Each design language carries very specific syntax and semantics [55]. The syntax of a language defines its grammar as a set of valid strings over an alphabet. While design languages are typically textual, some have an optional or exclusively graphical syntax (e.g. a flow chart as a graphical representation of an imperative program or the purely graphical StateCharts language). The semantics of a language subsequently defines the meaning of strings written in the language by mapping the syntax into an underlying semantic model, such as a mathematical domain [167] or an abstract state machine model [158, 82].

A description of a design in such a design language is then called a design model. When referring to models, we need to distinguish between a design model as an instance of a syntactically valid description written in the language, the semantic model underlying a language, and an MoC that defines a formal class of execution models where language-specific details such as data types and formats are abstracted away. For example, a MP3 decoder design model can be described as an instance of a KPN MoC captured in the syntax and semantics of the SystemC language. Typically, the same MoC can be represented in a variety of languages. Conversely, the same design language can represent different MoCs if it has a broad, basic semantic model that other MoCs map to. For example, while differing in their concrete syntax and detailed semantics, sequential programming languages such as C or C++ all support an imperative MoC, yet can also capture FSMs or FSMDs [184]. Note, however, that support for different MoCs in different languages varies, and specialized languages exist that are tied to a specific MoC, e.g., the graphical StateCharts language, which directly realizes a HCFSM model.

3.2.1   NETLISTS AND SCHEMATICS

Over the years, many new design languages have emerged to capture the necessary and sufficient semantics at each new level of abstraction. Early on, one of the first concepts that was formally modeled and captured was the notion of a netlist. Netlist models are purely structural representations of the design as a set of components and their connectivity. As such, netlist models are the basis for describing block diagrams used in early tools for computer-aided schematic entry and editing. Such schematic editors support simple automatic design rule checks to ensure, for example, that connections are only made between component ports that are compatible in terms of direction, signal and logic levels. Hence, their main role is documentation of block diagrams to facilitate communication between different design teams.

Some of the first design languages were developed for description of netlists at the gate level, such as the Electronic Design Interchange Format (EDIF). However, the concept of a netlist for structural representations is universal and has been carried over into corresponding new languages at each level. Today, at the system level, variants of the Extensible Markup Language (XML) are commonly used to capture netlists of system platforms. For example, the SPIRIT consortium defines the IP-XACT standard [172] for XML-based exchange and assembly of system-level IP components.

3.2.2   HARDWARE-DESCRIPTION LANGUAGES

After capturing netlists and schematics, interest arose in representing not only the structure of designs but also their design behavior. By adding capabilities for describing the behavior of every component in a netlist, languages gained execution semantics and could be simulated to validate the design. In order to remain general, most widely used design languages are based on a very basic discrete-event execution model. In a discrete event MoC, the system is represented as an ordered sequence of events where each event is a *(value, tag)* tuple that marks a change of state in the system at a certain point in simulated time [163]. Depending on the value and tag types supported by a specific language, events can be used to model arbitrary state changes. For example, a signal is defined as a sequence of voltage changes on a wire.

On the one hand, many different designs objects and classes of MoCs can be mapped to such a universal event model and represented in a single language with a small set of basic primitives. On the other hand, as described in Chapter 1, while this expressibility might be desirable for simulation purposes, there is a trade-off with the unambiguousness needed for analysis, synthesis and verification, typically restricting corresponding use of such general languages to a well-defined subset with well-defined and unique interpretation.

In the early stages, these ideas were applied to the description of hardware blocks at the gate level. Later on, they were transferred to the Register-Transfer Level (RTL). This resulted in the definition of so-called Hardware-Description Languages (HDLs) such as VHDL [7] or Verilog [180]. For example, at the RT level, the design is described as a microarchitecture consisting of functional and storage units connected by wires. Each RT component, such as a register or an ALU, will eventually consist of logic gates, while its behavior is inherently modeled in the form of Boolean expressions. Since logic gates and consequently RT units respond to the signal changes at their inputs over time, a mechanism is needed to indicate when inputs change and when this change propagates to the outputs. Therefore, an event-driven execution behavior was added to trigger evaluation of a component on every input change and subsequently propagate the new results to all components connected at the outputs. This process of event propagation and evaluation is repeatedly performed to simulate the design behavior over time. All together, event-driven execution allowed for the first time to completely simulate a digital circuit in the computer.

For simulation of HDL models, a so-called discrete event simulator internally maintains a logical simulated time and a queue of events ordered by their time stamps. In each simulation cycle, the simulator dequeues all events with the current time stamp and triggers execution of processes waiting for those events. Each component in the design is thus associated with a process describing its functionality. Once triggered, the process body is executed to compute internal state changes and a set of new values on its output signals. The simulator then inserts the events generated by each process into its queue and advances time to start the next simulation cycle. Note that to model delays, processes can post events at future points in time or can wait for time to advance in the middle of their execution. Furthermore, in more recent languages, processes can dynamically change their sensitivity and wait for and post arbitrary events throughout their execution. Hence, many HDLs can also model abstracted behavior of process networks beyond simple components connected only by wires.

3.2.3 SYSTEM-LEVEL DESIGN LANGUAGES

As we move to the system level, it becomes important not only to model the hardware side of a design, but also parts of the system implemented in software. As a result, new languages have been developed that add capabilities to describe software in a native manner. Due to the large body of legacy code, a natural choice is to include software in the form of standard C code, either by combining C with an existing HDL (SystemVerilog [174]) or by adding hardware modeling capabilities to C or C++ through extensions (SpecC [65]) or via libraries (SystemC [81]).

In all cases, and in the same way as previous HDLs, such System-Level Design Languages (SLDLs) are based on a discrete event driven execution model that supports necessary concepts for concurrency, hierarchy, timing and synchronization. In addition, SLDLs are supplemented with support for rich, abstract data types, process- and state-based computation, and libraries of communication channels. For example, the SpecC language implements a PSM MoC with C-based processes composed hierarchically in an arbitrary parallel, pipelined, sequential or state machine fashion. Furthermore, SpecC introduced the concept of native channels with a library providing, among others, message-passing, handshake, queue and semaphore type communication. SystemC, on the other hand, started out as a C++-based HDL with parallel processes and signals but later gained similar abstract channel concepts and libraries.

In addition, there are proprietary and standardized approaches that aim to provide metalanguages for formally capturing heterogeneous models including associated requirements and constraints [12, 5]. In all cases, such SLDLs allow us to describe complete systems and their applications within a single framework all the way from abstract specification of high-level MoCs down to processor implementations at the RT level.

3.3   SYSTEM MODELING

System design in general describes the process of going from a high-level system specification of the desired functionality down to a system implementation at the RT or instruction-set level. As outlined in Chapter 1, however, the semantic gap between specification and implementation is too large to be closed in a single step. Following a top-down or meet-in-the-middle approach, the system design process is therefore broken into a series of smaller steps. At the core of this process are definitions of system models to represent and pass design information from one step to the next.

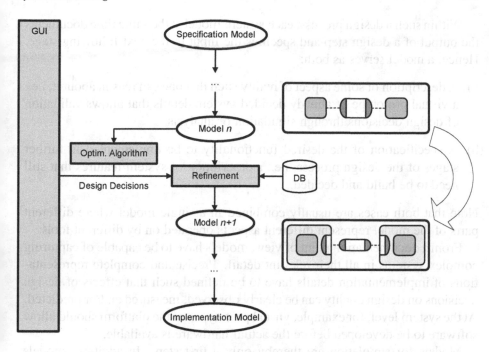

FIGURE 3.6 System design and modeling flow

3.3.1 DESIGN PROCESS

Realizing a Specify-Explore-Refine (SER) methodology based on model algebra principles described in Chapter 1, a general design flow can be established in which an initial specification is gradually brought down to a final implementation through successive, stepwise refinement of design models (Figure 3.6 [71]). In each design step, a refinement tool takes the input model and implements a set of design decisions in order to generate an output model at the next lower level of abstraction. In the process, tools insert a new layer of computation and/or communication detail that reflects and represents the given decisions. For example, to implement communication between software and hardware, refinement tools generate drivers and interrupt handlers inside a model of the processor hardware.

Design decisions can thus come from the designer, typically entered interactively through a Graphical User Interface (GUI), or from an automated algorithm. In both cases, decisions are made to optimize a set of design metrics, e.g., to simultaneously minimize system cost and area while not exceeding a maximum latency and power consumption. In general, both refinement and decision-making can be manual or automated.

Within such a design process, each system model at the same time documents the output of a design step and specifies the input to the next following stage. Hence, a model serves as both:

(a) A description of some aspect of reality such that one can reason about it, i.e., a virtual prototype of already decided system details that allows validation of design decisions through simulation or analysis.

(b) A specification of the desired functionality to be implemented in further stages of the design process, i.e., a description of system features that still need to be build and decided.

Note that both cases are usually combined in a single model where different parts of the model represent different aspects operated on by different tools.

From a documentation point of view, models have to be capable of capturing complex systems in all their relevant detail. Precise and complete representations of implementation details have to be defined such that effects of design decisions on design quality can be clearly observed, measured and/or predicted. At the system level, for example, virtual prototypes of the platform should allow software to be developed before the actual hardware is available.

Models for simulation are thereby only a first step. In addition, models should also enable formal methods to be applied for static analysis and verification of design properties, e.g., to guarantee response times in a hard real-time environment. Especially also under the aspect of serving as a specification for further synthesis, models have to be defined with unambiguous semantics, such that application of corresponding tools becomes possible in the first place.

Combining documentation and specification aspects, each design model is an abstracted representation of a design instance. A model is associated with a corresponding abstraction level that defines the granularity of implementation detail represented in the model. As design progresses, we gradually move down in the level of abstraction by adding more and more implementation detail. Due to the lack of detail at higher levels, models simulate faster, but the accuracy of results is limited, typically resulting in a trade-off as we move up in abstraction. Therefore, the ideal design process should support a variety of levels with different trade-offs, both to break the design flow into smaller steps and for efficient design space exploration. For example, in the early stages, designers want to rapidly prune the design space of clearly infeasible solutions. As such, early models have to be fast but accurate only in relative, not absolute, terms. Then, as the design space continues to shrink, we can gradually afford to spend more and more time on slower but increasingly accurate simulation until a final solution is confirmed.

3.3.2 ABSTRACTION LEVELS

As described in Chapter 1 and Chapter 2, there are four main abstraction levels representing circuit, logic, processor and system levels. Within each level, there are many different implementation details to be considered and design steps to be performed. This requires main abstraction levels to be divided into several intermediate levels and corresponding design models to be defined for each. Specifically, in relation to system design we need to be concerned with the implementation details within the upper system and processor levels.

Following the separation of computation and communication, we can first and foremost distinguish between largely orthogonal computation and communication details [107, 72]. Both can range from purely functional, fully untimed descriptions down to cycle-accurate levels. On the communication side, transaction-level approaches have recently become popular as an intermediate approach for modeling of communication detail above the cycle-accurate level [31, 77, 109]. Similar concepts can also be applied on the computation side.

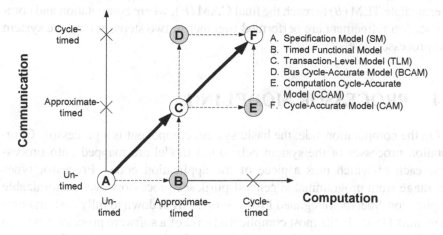

FIGURE 3.7 Model granularities

Figure 3.7 [31] shows the range of granularities of computation and communication for various levels of abstraction. For example, a behavioral system specification at the origin of the graph (*A*) is untimed in both computation and communication with only a causal ordering between processes. Annotating computation with execution models and estimated or measured delays results in a timed functional model (*B*). By also further refining communication down to timing-accurate bus transactions, we reach a Transaction-Level Model (TLM) of the system at point *C*. As such, a TLM includes timed models of computation and communication behavior for the processors in the system.

Going into the design process at the processor level, Bus-Functional Models (BFMs) of processors (which are assembled into a Bus Cycle-Accurate Model (BCAM) of the system) can be obtained by refining interfaces down to state machines that drive and sample bus wires on a cycle-by-cycle basis (*D*). Alternatively, implementing only computation in the processors down to a microarchitecture at the RT or timed instruction-set level, leads to a Computation Cycle-Accurate Model (CCAM), shown as point *E* in the graph. Finally, the combined Cycle-Accurate Model (CAM) at the lowest level (*F*) is cycle-timed in both computation and communication.

As mentioned before, a methodology is defined as a set of models and transformations in between. A specific system design methodology is then established through the path that is taken to go from an untimed specification at point *A* all the way to a final cycle-accurate implementation at point *F*. To that effect, the path taken determines the intermediate system models that are available, as well as the amount, type and order of refinements to be performed throughout the design process. For example, in a general design methodology (see Chapter 1), design starts with a specification model (*A*) and progresses through an intermediate TLM (*B*) to reach the final CAM (*F*), where computation and communication refinement are performed together in two steps, at both the system and processor levels.

3.4 PROCESSOR MODELING

On the computation side, the basic system component is a processor. Computation processes of the system behavioral model are mapped onto processors, each of which runs a piece of the application code. Processor types can range from programmable general-purposes processors over customizable Application-Specific Integrated Processors (ASIPs) down to fully custom hardware units [184]. In the most complicated case of a software processor, we can typically distinguish several layers of computation implementation (Figure 3.8). In the embedded case, the behavior of the system and hence of its components is not only defined by their functionality but also, of equal importance, by their timing. Thus, both aspects have to be considered throughout the design process. Since all of these layers significantly contribute to the functional or, more importantly, overall timing behavior of a processor, an accurate processor model has to include all of them [166, 24].

At the specification level, the application is modeled as a network of communicating processes. Inside the processes, basic application algorithms are typically described in an imperative programming language such as C. However, application code in itself is untimed. To introduce the notion of time, information about execution delays on the given target processor has to be inserted into the code. This back-annotation can be performed at varying levels of granularity. Such back-annotated application code then has to be augmented with a model of its execution environment such that effects of running the code on a given platform are accurately described.

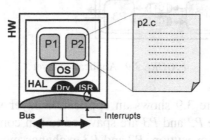

FIGURE 3.8 Processor modeling layers

In the case of a software processor, as shown in Figure 3.8, application processes usually run on top of an operating system (OS), which provides dynamic scheduling and multi-tasking services. On the other end, application and OS software has to run on top of the actual processor hardware, which realizes physical bus interfaces and interrupts (including processor suspension and interrupt timing) for communication with the external world. In between, a hardware abstraction layer (HAL) provides canonical interfaces, such as bus drivers and interrupt service routines (ISRs) for accessing the processor hardware from the software (i.e., application and OS) side.

Note that in general, all layers together determine the final execution order and have a large influence on the overall timing behavior of the application running in the processor. Hence, models of all layers and their relevant details need to be developed and integrated into the design process. To that effect, Figure 3.8 shows the most general case of a software processor with all layers. By contrast, models of other processors, such as custom hardware units, are derived as specialized versions of this general model by not including OS or hardware abstraction layers.

3.4.1   APPLICATION LAYER

As mentioned previously, at the highest application layer, computation functionality running on a processor is generally described as a hierarchical set of communicating processes. Processes at the leaves of the hierarchy encapsulate basic algorithms, e.g., in the form of standard ANSI C code. Processes can be composed hierarchically in an arbitrary serial-parallel fashion. Furthermore, processes communicate via shared variables, events or abstract channels providing high-level, typed communication primitives such as message-passing, queues or semaphores.

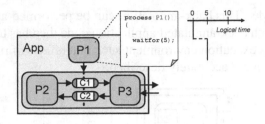

FIGURE 3.9 Application layer

For example, Figure 3.9 shows an application layer with three processes *P1*, *P2* and *P3*, where *P2* and *P3* are spawned to run concurrently after *P1* is finished. During their execution, *P2* and *P3* exchange messages over channels *C1* and *C2*. Furthermore, *P3* communicates with other processors in the system through two external ports.

To provide the necessary concepts of concurrency, communication and timing, such application descriptions are typically modeled in a C-based SLDL such as SystemC or SpecC. With such an application specification, the designer is essentially provided with a high-level, abstract model for programming the complete platform across different processors.

To achieve desirable high simulation speeds, the application model is executed on the event-driven SLDL simulation kernel running natively on the simulation host. In the process, application code is compiled into native host instructions to directly emulate its functional behavior using the fastest possible host execution. In order to provide additional feedback about the timing behavior, application processes have to be back-annotated with execution timing, which models and simulates the delays of running application code on the chosen target processor in the final design (see Chapter 4).

As shown in Figure 3.9, back-annotation is performed by inserting wait-for-time statements into the code as supported by the timing model of the underlying SLDL. Depending on the available data and the use case, such `waitfor` statements can be inserted at different levels of granularity ranging from basic blocks up to the level of functions or whole processes. Each option thereby results in a specific speed/accuracy trade-off. For example, back-annotation at the function or process level cannot represent dynamic effects of data-dependent control flow in the code. Instead, whole blocks of code are associated with a single, static delay number based on worst-case or average-case assumptions. On the other hand, back-annotation at the basic block level accurately models control flow dependencies but is slower due to the larger number of `waitfor` statements to be simulated.

There is a multitude of sources for obtaining execution delay numbers to be back-annotated into the application code. In general, delays can be acquired either through estimation or measurement. Estimation of execution time is a topic that has been studied extensively with approaches that range from purely static, worst-case estimation techniques [194] to profiling-based solutions [31] that aim to deduce estimates from analysis or simulation of the source code, respectively. In terms of delay measurements, cycle counts for the code can be gained by compiling the code for the target processor and tracing its execution either on the real processor or in a corresponding timing-accurate ISS. Lastly, hybrid approaches exist that compile code down to an intermediate level in order to apply analysis and estimation techniques at a level closer to the implementation [93].

3.4.2   OPERATING SYSTEM LAYER

In the application layer, computation processes are modeled as running truly concurrently. In reality, however, we have to assume that processors can only execute a single thread of control or a limited number of threads at any given time. With the operating system layer, the goal is therefore to introduce accurate representations of the scheduling of parallel processes on the inherently sequential processors.

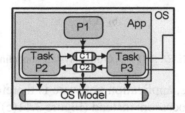

FIGURE 3.10 Operating system layer

As a first step, processes are grouped into tasks where all processes within a task are arranged in a fixed order according to a pre-defined static schedule. As shown in Figure 3.10, for example, processes *P2* and *P3* are converted into tasks with static scheduling combining all sub-processes into one sequential piece of code each. In a second step, remaining tasks are then considered to be dynamically scheduled during runtime, typically by a Real-Time Operating System (RTOS). To accurately reflect and specify these dynamic scheduling and RTOS effects and needs, an abstracted model of the RTOS is inserted into the processor's operating system layer [75]. In the process, tasks (e.g., *P2* and *P3* in Figure 3.10) are refined to run on top of the OS model by inserting the necessary OS calls for task management (creation and deletion), synchronization (event handling) and timing (delay modeling). In addition, existing application channels (e.g., *C1* and *C2*, Figure 3.10) are refined into a model of Inter-Process Communication (IPC) that is properly integrated with the OS model by inserting appropriate OS calls for implementation of synchronization.

The operating system layer and OS model describe expected RTOS behavior, including the desired scheduling algorithm and scheduling parameters, both for validation during simulation as well as for further synthesis. To that effect, the concept of OS modeling in general is based on the idea that at the specification level, as shown in Figure 3.11(a), processes are executed directly on the underlying simulation kernel. Simulations run at native speeds but the concurrency model of the simulator does not match the actual scheduling algorithm implemented in the real RTOS.

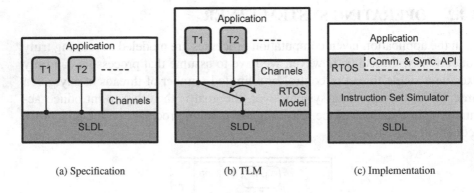

(a) Specification   (b) TLM   (c) Implementation

FIGURE 3.11 Operating system modeling

To get accurate results, application software is therefore traditionally simulated in ISS models of processors instead (Figure 3.11(c)). For this purpose, the application code is cross-compiled for the target processor and linked against the real target RTOS libraries. The resulting final target binary is then executed by an ISS, which in turn can be integrated into an SLDL environment for co-simulation with the rest of the system [20, 70]. Such ISS approaches can be very accurate, but as a result of their accuracy, can also be slow, especially if multiple processors have to be co-simulated together in a cycle by cycle fashion.

The goal of high-level RTOS modeling is thus to provide a solution that combines the speed of native application execution (Figure 3.11(a)) with the accuracy of an ISS model (Figure 3.11(c)). Instead of running the real operating system with all of its associated overhead, an abstracted RTOS model is inserted as an additional layer that sits between the application and the underlying simulation kernel (Figure 3.11(b)). The OS model abstracts away unnecessary implementation details and focuses solely on modeling key concepts relating to multi-tasking, preemption, interrupt handling and inter-process communication and synchronization. As such, the RTOS model adds only a negligible simulation overhead. On the other hand, it provides accurate feedback about all important OS effects early on in the design process and at a high level of abstraction.

Internally, the OS model wraps around and replaces the underlying SLDL event handling with its own primitives. As mentioned above, tasks and channels are refined to call the equivalent OS model services and are not allowed to access the simulation kernel directly. Instead, the OS model selectively relays calls to the kernel, ensuring that at any given time only one task is active and all other tasks are blocked at the SLDL level. Whenever the OS model is called, either by a task, a channel or from an asynchronous event such as an interrupt handler, a re-scheduling and task switch is triggered. The OS model then blocks the current task and selects, dispatches and releases a new task based on its internally re-implemented scheduling algorithm. For example, an OS model that emulates a priority-based scheduling will block a low-priority task calling the OS if in the meantime a higher priority task has been activated by an asynchronous external interrupt. By replacing delay models (i.e., `waitfor` statements) with an appropriate wrapper, an OS model can therefore simulate task preemption accurately within the granularity of the given back-annotation.

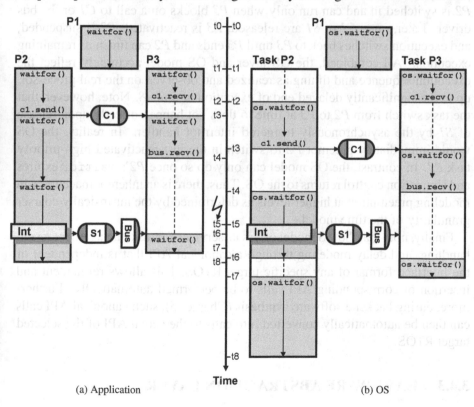

(a) Application   **Time**   (b) OS

FIGURE 3.12 Task scheduling

Figure 3.12 shows the resulting execution schedules for the example previously introduced in Figure 3.9 and Figure 3.10. Execution starts at time *t0* with process *P1*, which in turn spawns processes *P2* and *P3* at time *t1*. In the unscheduled case (Figure 3.12(a)), *P2* and *P3* are running truly concurrently and their simulated execution times overlap unless there is some causal dependency between them. For example, at time *t2*, process *P3* is blocked and waits for a message from *P2*, which arrives over channel *C1* at time *t3*. Similarly, at time *t4*, *P3* enters the bus driver to wait for external data from another processor in the system. At time *t5*, an external interrupt to signal availability of data arrives. The corresponding interrupt service routine is executed and releases a semaphore *S1* in the bus driver. The driver in turn receives the external data and finally resumes execution of *P3* at time *t6*. All throughout, *P2*, on the other hand, runs continuously, uninterrupted by any of the events in the system.

Once the OS model has been inserted (Figure 3.12(a)), execution of tasks *P2* and *P3* is interleaved according to the selected scheduling algorithm. In this example, priority-based scheduling is employed where task *P3* has a higher priority than task *P2*. Hence, *P3* executes unless it is waiting on some event. *P2* is switched in and can run only when *P3* blocks on a call to *C1* or the bus driver. Later, once *C1* or *S1* are released, *P3* is reactivated, *P2* is suspended, and execution switches back to *P3* until *P3* ends and *P2* can finish its remaining execution. All combined, the OS layer and OS model accurately reflect the execution sequence and timing as realized and observed on the real processor, up to the significantly delayed end of execution at time *t8*. Note, however, that the task switch from *P2* to *P3* at time *t6* does not happen directly upon release of *S1* by the asynchronously triggered interrupt handler. In reality, the OS would immediately preempt *P2* to switch in the just reactivated high-priority task *P3*. In contrast, the OS model can only do so once *P2*'s `waitfor` expires and simulation control returns to the OS. Thus, there is an inherent inaccuracy in modeling preemption at higher levels as determined by the intrinsically coarser granularity of the time model.

Finally, note that the OS model offers its services for task management, event handling, and delay modeling through a canonical API that is independent of the interface format of any specific target RTOS. This allows refinement and insertion of corresponding API calls to be performed automatically. Furthermore, during backend software synthesis (Chapter 5), such canonical API calls can then be automatically converted into calls to the actual API of the selected target RTOS.

3.4.3   HARDWARE ABSTRACTION LAYER

The Hardware Abstraction Layer (HAL) provides the lowest level of functionality that is implemented in software. As such, its border marks the boundary between the processor hardware and the software running on top of it. In other words, all layers thus far down to and including the HAL are implemented in software whereas all layers outside of the HAL are realized in hardware. Again, this is for the general case of a software processor. In the case of a custom hardware unit, neither HAL nor OS layers exist, and all functionality is implemented in hardware.

Basic HAL templates are typically stored in the PE database, containing pre-defined functionality for abstracting access to the processor hardware into a set of canonical interfaces and services. For example, a HAL model in the database will include the hardware-specific driver code for transferring arbitrary blocks of bytes over the processor bus interface. Likewise, the HAL contains templates of low-level interrupt handlers that are properly associated with corresponding hardware interrupt sources.

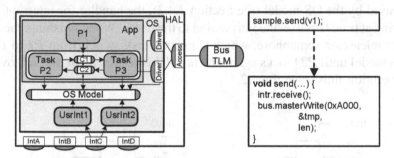

FIGURE 3.13 Hardware abstraction layer

On top of the HAL templates stored in the database, models of the drivers for communication with the external world can then be developed and inserted. Details of these communication models will be described in Section 3.5. In general, as shown in Figure 3.13, models of the drivers are inserted into the OS layer, where they are integrated with the rest of the processor model, i.e., the OS model and the low-level bus access code in the HAL. In the process, low-level interrupt handler templates in the HAL (e.g., *IntA* through *IntC* in Figure 3.13) are filled with required code. Furthermore, necessary user-level interrupt tasks (*UsrInt1* and *UsrInt2*) are generated and registered with low-level handlers and the OS model.

Together, these drivers describe the implementation of application channel calls (such as send() or receive()) down to the level of external interrupts and bus read() or write() transactions. As a result, HAL models of processors are connected by and communicate with each other through bus models at the level of individual interrupts and bus transactions. As mentioned previously and as will be described in more detail in Section 3.5, such bus TLMs provide an abstracted implementation of bus communication beyond individual pins and wires.

3.4.4 HARDWARE LAYER

With the final hardware layer, an accurate model of the actual processor hardware is included. The processor hardware model specifically captures details of physical bus interfaces and of interrupt handling behavior to suspend regular execution and handle exceptions whenever an external interrupt is received. Note that in the HAL, interrupts are handled concurrently to the application and OS code, and the execution of interrupt handlers overlaps with those of the tasks. For example, Figure 3.14(a) shows the schedule of execution for the HAL model from Figure 3.13. While tasks *P1* and *P2* are serialized and interleaved by the OS model (see Section 3.4.2), the handler for interrupt *IntC* occurring at time *t1* is executed in parallel to the tasks. When finishing, the *IntC* handler releases a semaphore, which triggers a task switch from *P1* to *P2* in the OS model until *P2* blocks again and task *P1* is resumed. In the end, overall task execution finishes at time *t2*.

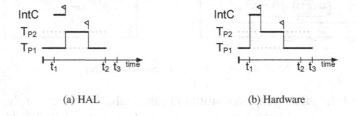

(a) HAL   (b) Hardware

FIGURE 3.14   Interrupt scheduling

In reality, however, the processor hardware interleaves interrupt processing with execution of regular code. As shown in Figure 3.14(a), when the interrupt occurs at time *t1*, the processor hardware suspends execution of all tasks to execute the *IntC* handler in its own context. Only after *IntC* is finished will the hardware resume task execution, at which point the OS can execute the interrupt-triggered switch from *P1* to *P2*. As a result, end of task execution in the real processor is delayed until time *t3* instead of *t2*. Since interrupt processing and the interrupt load on the processor can therefore have a significant effect on overall timing, corresponding models need to be included and considered in the design process [198].

For this purpose, the hardware layer as shown in Figure 3.15 adds a separate model of the processor's hardware interrupt logic. When receiving an external event on one of its external interrupt ports (*INTA* through *INTC*, for the example

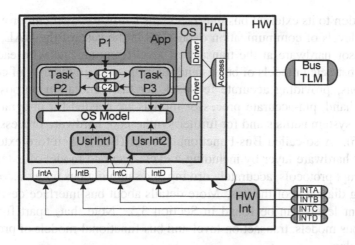

FIGURE 3.15 Hardware layer

in Figure 3.15), the hardware interrupt logic suspends execution of the complete processor software including HAL, OS and application layers. Replicating the processor's behavior for interrupt nesting and interrupt priorities, the interrupt logic then calls the appropriate interrupt handler (e.g. *IntC* in Figure 3.15). The handler in turn notifies a user-level interrupt task, e.g., *UsrInt1* or *UsrInt2*, and returns. Upon return of the interrupt handler, the HW interrupt logic resumes execution of HAL, OS and application software, and the OS model re-schedules tasks such that interrupt-triggered high-priority jobs (e.g., *UsrInt1* or *UsrInt2*) can potentially preempt any currently running task (e.g., *P2* or *P3*).

Note that even though interrupt handlers behave in essentially the same fashion as high-priority tasks, scheduling strategies for tasks and interrupts in general are different, as is the case for a round-robin OS that runs in a processor with prioritized interrupt sources. Therefore, interrupt handling behavior has to be modeled separately. In essence, hardware and OS layers implement a hierarchal scheduler in which a primary interrupt scheduler in the hardware exercises control over the secondary OS-internal task scheduler, each with its own respective scheduling strategy that accurately represents reality.

Next to the core processor suspension functionality, the HW layer also includes any interrupt controllers and other peripherals (e.g., timers) immediately associated with the processor. The interrupt controller together with the suspension logic then determines the overall interrupt behavior and interrupt scheduling strategy. If interrupt controllers are programmable, the HAL will include the necessary code for setting up and properly associating interrupt handlers with external interrupt sources according to user-selected parameters such as priorities.

In relation to its external bus interfaces, the hardware layer can provide two different levels of communication detail. As in the case of the HAL, models of processor hardware at the transaction level communicate with each other through abstracted models of busses and interrupts at a granularity of complete transactions, providing accurate feedback for fast simulation purposes. On the other hand, pin-accurate processor models are needed for integration into structural system netlists and for further synthesis of hardware processors (see Chapter 6). A so-called Bus-Functional Model (BFM) therefore extends the processor hardware layer by including a cycle-accurate model of external bus and interrupt protocols, accurately driving and sampling bus wires according to the timing diagrams of the bus. More details about bus interface descriptions at different levels can be found in Section 3.5. Note that, apart from their external bus models, transaction-level and bus-functional models of processors are equivalent.

TABLE 3.1 Processor models

Features		
Target approx. computation timing	Appl.	
Task mapping, dynamic scheduling		OS
Task communication, synchronization		HAL
Interrupt handlers, drivers		HW
HW interrupt handling, int. scheduling		ISS
Cycle-accurate computation		

In summary (Table 3.1 [166]), a model of system computation as implemented by custom or standard processors can be constructed in a layer-based fashion. With each layer, new features and aspects of computation behavior are specified and can hence be observed. Basic algorithmic functionality, including execution timing, is modeled at the application layer. The operating system layer adds dynamic scheduling and OS effects. At the hardware abstraction layer, interrupt handlers and bus drivers for external communication are inserted. Finally, the hardware layer adds a model of the processor hardware and interrupt logic.

In the end, a processor model at the hardware level provides a complete and full-featured description of computation running in its execution environment. Transaction level versions of hardware processor models serve as a fast yet accurate virtual prototypes for simulation in close to real time. Bus-functional processor models, on the other hand, are structurally accurate for system integration and hardware synthesis. As a reference, all such high-level processor models are compared against a traditional bus-functional ISS model of the same processor (see Chapter 5), which can be fully cycle accurate at the expense of slow simulation speeds.

3.5 COMMUNICATION MODELING

Communication needs and principles have been studied extensively in the general-purpose networking community. The networking community early on developed the ISO/OSI 7-layer model [98] to reason about networks and to serve as a guideline and outline of requirements and implementations. The model divides the functionality generally required to implement any network into seven different layers. Layers are grouped and ordered based on orthogonality. Each layer defines the functionality implemented by it and the semantics supported at its interface. Layers are stacked on top of each other and each layer provides services to layers above by using services of the layer below. As such, the model provides a clear separation of concerns and allows for a structured process when designing new networks.

The ISO/OSI model has proven to be very valuable for reasoning about the design process in the networking world. Based on the observation that system communication is not inherently different from any general-purpose communication, we can adopt the ISO/OSI model as a basis for organizing system communication functionality, the communication design process, and communication modeling. In doing so, however, layers need to be tailored to specific system design requirements. For example, lower media access layers have to be split in order to reflect the separation between functionality implemented in hardware and software. Furthermore, layering has to take into account the special features and restrictions of on-chip and off-chip busses employed in the embedded world. Note that in the context of this book, the term bus is used broadly in the sense that it can refer to a wide variety of physical communication media, including, for example, serial, point-to-point or network-oriented busses such as RS232, CAN, Ethernet or wireless protocols.

Table 3.2 [74] summarizes the results of this process. Next to related original ISO/OSI numbers, the table lists for each layer semantics provided at the interface to the layer above, functionality realized inside, and corresponding processor layers into which this functionality is eventually inserted. Note that presentation, session, transport and network layers form the network-level drivers that are implemented next to the OS. Link, stream, and media access layers, on the other hand, are implemented as low-level drivers and interrupt handlers directly on top of the processor HAL. Finally, the protocol layer implements the actual physical bus interface in the processor hardware.

TABLE 3.2 Communication layers

Layer	Semantics	Functionality	Implementation	OSI
Application	Channels, variables	Computation	Application	7
Presentation	End-to-end typed messages	Data formatting	OS	6
Session	End-to-end untyped messages	Synchronization, multiplexing	OS	5
Transport	End-to-end data streams	Packeting, flow control	OS	4
Network	End-to-end data packets	Subnet bridging, routing	OS	3
Link	Point-to-point logical links	Station typing, synchronization	Driver	2b
Stream	Point-to-point control/data streams	Multiplexing, addressing	Driver	2b
Media access	Shared medium byte streams	Data slicing, arbitration	HAL	2a
Protocol	Media (word/frame) transactions	Protocol timing	Hardware	2a
Physical	Pins, wires	Driving, sampling	Interconnect	1

Taken as a whole, Table 3.2 outlines the needs and requirements for implementation of system communication. In order to support the design process for arbitrary communication architectures, we need to develop models of all communication layers and their relevant functionality. Details of layers and such models will be discussed in the following sections.

It should be noted, however, that the ISO/OSI model was only ever intended to aid in reasoning about communication stacks and their requirements. It was not meant to provide a reference for implementation of such functionality in the same layered manner. Thus, while layers serve as a basis for development and organization of communication models, any implementation, and hence any further synthesis of models, should consider the complete stack as a single specification for possible merging of functionality and optimization across layers.

3.5.1   APPLICATION LAYER

As discussed previously for the computation side, the application layer provides the designer with a high-level programming model for describing system behavior. In this model communication semantics can vary widely in terms of required or desired data and control flow. At the application level, a rich and powerful set of high-level communication semantics should therefore be supported. Typical examples of commonly used application-level communication primitives include:

(a) Pure events that establish one-way synchronization (i.e., control flow transitions) but do not carry data.

(b) Shared variables, which only hold data and do not include any synchronization.

(c) Synchronous and asynchronous message-passing channels, which combine data transfers with, respectively, two-way or one-way synchronization (control flow). In the synchronous case, sender and receiver block until both are available to exchange data. In the asynchronous case, only the receiver side is blocking, waiting if necessary for data from the sender to become available. The sender, on the other hand, may or may not block. Asynchronous implementations can use buffers to decouple the sender from the receiver, such that a sender is independent and will only block if no buffer space is available. The amount of buffering, however, is implementation-dependent and, in general, no guarantees are made on the sender side. Note that in an extreme case, asynchronous message-passing can be implemented synchronously without any buffering.

(d) Queues as a special case of asynchronous message-passing with well-defined, fixed buffer sizes to implement and guarantee a specific queue depth.

(e) Complex and user-defined channels with extended semantics, such as semaphores or mutexes widely used for advanced synchronization in the software world.

In all cases, application designers expect communication primitives to provide guaranteed delivery. As such, all provided and supported primitives are assumed to be reliable, lossless and error-free. Communication design thus needs to ensure that these primitives are implementable on any given target communication architecture, even if underlying physical communication media are unreliable.

In general, a system specification can contain communication in the form of any of the application-level primitives described above. In contrast, actual implementations of communication stacks can typically only support system-level communication using a canonical, restricted set of primitives. In the course of mapping system behavior onto processors, the application layer therefore has to translate all communication between system components into primitives supported by the following design process.

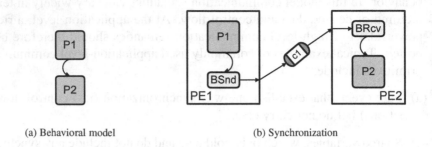

(a) Behavioral model   (b) Synchronization

FIGURE 3.16   Application layer synchronization

To begin with, the application layer has to insert any necessary synchronization to guarantee that original execution semantics are preserved. Consider,
for example, a system behavior in which two processes *P1* and *P2* are composed sequentially (Figure 3.16(a)). Mapping *P1* and *P2* to *PE1* and *PE2*,
respectively, results in them running in parallel on the inherently concurrent
processors. Thus, the implicit sequential transition from *P1* to *P2* has to be
converted into explicit synchronization between the PEs. As shown in Figure 3.16(b), an additional process *BSnd* on *PE1* transfers control to *PE2* by
sending a message over a newly inserted channel *C1* as soon as *B1* finishes.
Process *BRcv* on *PE2*, on the other hand, ensures that *B2* does not start until the
synchronization message on *C1* has been received and hence, as desired, until
B1 on *PE1* has finished.

In addition to transitions, the system behavior might contain shared variables
between processes. If, as shown in Figure 3.17(a), two processes *P1* and *P2*
are mapped to different processors, storage represented by the shared variable
v1 in between needs to be moved into local memories of one or more system
components. In a distributed implementation (Figure 3.17(b)), a local copy of *v1*
is created in each accessing PE. Local copies are kept synchronized by inserting
messages to exchange updated variable values at synchronization points. For
example, in Figure 3.17(b), updates of *v1* are merged into previously generated
synchronization messages. Together with passing control from *P1* on *PE1* to
P2 on *PE2*, *BSnd* and *BRcv* hence transfer any necessary updates of *v1* via
channel *C1*.

In other implementations, variables are mapped specifically into a chosen
single PE. In a memory-mapped I/O implementation, variables are mapped
into the local memory of a processor, typically one of the processors accessing the variable. For example, in Figure 3.17(c), *v1* is mapped into the *HW*
PE where it becomes a local register. A memory interface providing external
methods for reading and writing the register is then created in the *HW* such
that the *CPU* and other PEs can access *v1* over the network. Alternatively, in a

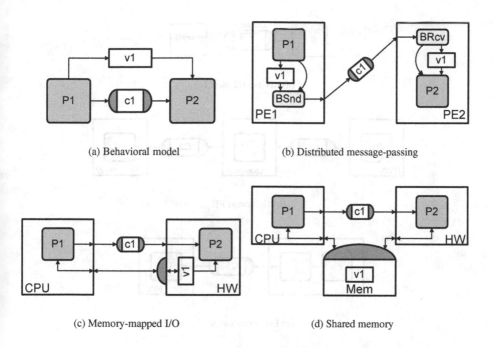

(a) Behavioral model   (b) Distributed message-passing

(c) Memory-mapped I/O   (d) Shared memory

FIGURE 3.17   Application layer storage

shared memory implementation (Figure 3.17(d)), *v1* is mapped into a dedicated memory component. Again, the memory provides an interface for other PEs (i.e., *CPU* and *HW*) to randomly access any variables stored inside.

Similar to shared variables, complex channels supported at the application level (such as queues, semaphores or mutexes) might have storage or computation requirements as part of their implementation. As such, they have to be similarly resolved into basic channel primitives supported by the following design process. For example, Figure 3.18(a) shows two processes, *P1* and *P2*, communicating through a queue channel, which requires both storage and computation to manage internal buffers and realize its external enqueuing and dequeueing functionality. As part of the application layer, such channels are translated into a client-server implementation. A separate server process is created to implement channel storage and functionality. Accesses to the channel by other client processes are translated into a Remote Procedure Call (RPC) to the server that emulates the original channel interface and semantics. Figure 3.18(b) shows an example where the additional server process is mapped into a dedicated, separate *Queue* hardware. Processes *P1* and *P2* in *HW1* and *HW2* then access the *Queue* via basic message-passing channels *C1* and *C2*, respectively. On the other hand, the server process can be mapped into one of the accessing PEs instead. For example, in Figure 3.18(c), the queue process is

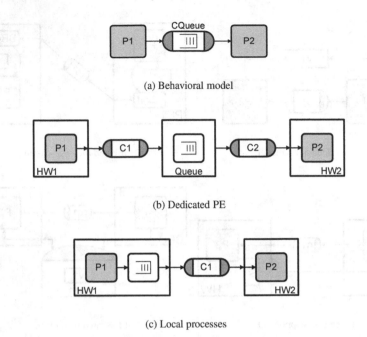

(a) Behavioral model

(b) Dedicated PE

(c) Local processes

FIGURE 3.18   Application layer channels

mapped into *HW1*, where it becomes an additional local process next to *P1*. In
this case, the server communicates locally with *P1* and accepts external requests
for read accesses from *P2* on *HW2*, again over a basic message-passing channel
C1.

After resolving synchronization, storage and complex channel requirements,
communication at the output of the application layer is reduced down to basic
message-passing and memory access primitives. This canonical set of primitives can then be further implemented on a given target communication architecture by the following layers of the communication stack.

3.5.2   PRESENTATION LAYER

The presentation layer is responsible for formatting of data and for conversion
of abstract datatypes found in the application into untyped blocks of bytes to be
transferred over the network. Generally, data is stored in different layouts in the
local memories of the different PEs. Layout of data items in a PE is dependent
on the PE's architecture and the target compiler/synthesizer used to translate
application code into a PE-specific implementation. The data layout of a PE is
sufficiently described by the following parameters:

(a) Bitwidth of a machine character, i.e., of the smallest addressable unit.

(b) Size (in machine characters) of each basic/primitive datatype.

(c) Alignment (in machine characters) of each basic/primitive datatype.

(d) Endianess (little or big).

Given this information, the exact layout of data in the PE's memory can be derived for all primitive or complex data types, such as arbitrary combinations of structures and arrays.

When exchanging data over the network, a common format of the data on the network has to be chosen for each pair of communicating partners. Similarly, a common layout of data in a shared memory accessed by two PEs has to be defined. In general, the format of data on the network or in the memory is equally defined by a set of character width, endianess, size and alignment parameters.

The data layout on the network or in the memory can be the same as in both, one or none of the involved PEs. If any network or memory parameter is different from the PE's characteristics, the presentation layer has to perform the necessary conversion for all accesses. By contrast, the presentations layer is empty and data can be copied one-to-one only if all parameters are the same.

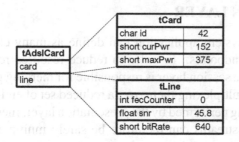

(a) Application data structure

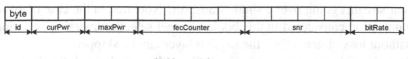

(b) Network byte stream

FIGURE 3.19 Presentation layer

Figure 3.19 shows an example of an application-level abstract data structure (Figure 3.19(a)) being marshaled into a stream of bytes to be transferred over the network (Figure 3.19(b)). At the application level, the *tAdslCard* data structure consists of two members (*card* and *line*) which are of *tCard* and *tLine* type, each containing several items of basic character, integer and floating-point data type. In the local PE memory, structures and their members are laid out to conform to PE parameters such as the requirement to align all items on word or double-word boundaries. In contrast, on the network level, the presentation layer transmits items byte by byte in the specified order and with given sizes (1 byte per char, 2 bytes per short and 4 bytes per int or float for this example), excluding in this case any empty bytes used for padding and alignment in the PE memory.

Note that the layout of data on the network or in memories can either be globally defined for the whole system or adapted differently for each pair of communicating PEs. In the former case, the layout can be optimized to reduce overall data traffic globally. For example, traffic is minimized by setting alignment and size parameters such that no byte padding is performed and all data types only occupy their minimally necessary space. In the latter case, the layout can be matched individually to one or both of the PEs, simplifying or completely avoiding presentation layer implementations.

All in all, the presentation layer translates typed messages and memory accesses between different endpoints at the application level into untyped messages and byte-wise accesses to be implemented by the layers below.

3.5.3   SESSION LAYER

In general, the system application can define as many channels as desired between any two endpoints. In order to reduce resource requirements in the following layers, the session layer is responsible for merging groups of channels into sessions and multiplexing them over a reduced set of end-to-end transports. After data formatting performed by the presentation layer, messages are untyped and resulting byte streams can therefore be safely multiplexed over a single transport channel. As such, the session layer is responsible for selecting the end-to-end channel used to transport messages of each application stream and to implement all the necessary means to separate and distinguish messages of different streams going over a single transport. Note that in the case of system communication considered in this book, channel selection/merging is optional and without loss of generality, the session layer can be skipped.

If two channels are accessed sequentially with a pre-defined, fixed and non-overlapping sequence of messages in both communication endpoints, they can be merged unconditionally into a single stream. Figure 3.20(a), for example, shows a case where two sequential processes (*P1* on the *CPU* and *P2* in *HW2*) communicate over two channels, *C1* and *C2*. Since processes can only exchange messages over one channel at a time, the session layer merges them by simply connecting both ports on both processes to a single shared channel

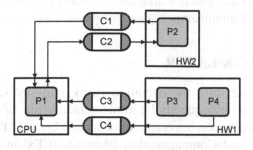

(a) Application channels

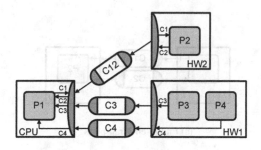

(b) Network transports

FIGURE 3.20   Session layer

C12 in between (Figure 3.20(a)). On both sides, message are then identified as originally belonging to either *C1* or *C2* simply by their order in the sequence of messages going back and forth over *C12* (i.e., the first message always being *C1* and the second *C2*).

In the case of concurrently accessed channels, messages cannot be identified by their order in the stream. For example, in Figure 3.20, channels *C3* and *C4* are fed by two concurrent processes *P3* and *P4* in *HW1*. As a result, messages may arrive in any order or may even overlap on the *CPU* side. Hence, when transferred over a single channel, messages originating from *C3* or *C4* cannot be directly distinguished. Instead, such concurrent channels can only be merged if additional headers are prepended to each message in order to identify messages belonging to different sources. In an embedded implementation, however, the additional overhead for handling and transmission of message headers is typically not justified and any such addressing realization is better deferred to lower layers (see Section 3.5.8).

In the end, the session layer handles the merging of channels by transmitting messages from different sources over a reduced set of end-to-end streams such that overall complexities and resource requirements in implementation of later layers below will be minimized.

3.5.4   NETWORK LAYER

The network layer splits the overall network into smaller subnets where different subnets can subsequently be implemented using different underlying media. In order to connect subnets and route packets between them, the network layer inserts additional Communication Elements (CEs) in between. In the process, the network layer transforms end-to-end transports of upper layers into logical point-to-point links between individual stations forming the overall network structure.

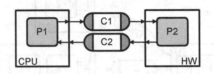

(a) End-to-end transports

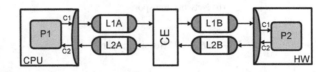

(b) Point-to-point links

FIGURE 3.21   Network layer

For example, in Figure 3.21, end-to-end channels *C1* and *C2* between *CPU* and *HW* PEs (Figure 3.21(a)) are each split into two logical links, *L1A/L1B* and *L2A/L2B*, respectively. In between, a *CE* routes and translates between links such that subnets on either side can each be implemented over different underlying busses.

We can generally distinguish two types of CEs, transducers and bridges, depending on the layer at which their functionality is realized. As shown in Figure 3.22(a), a transducer operates according to a store-and-forward principle. It receives blocks of data on one side (*PE1*), temporarily stores them in an internal buffer and does not send anything out on the other end (*PE2*) until the complete packet has been received and buffered. As a result, transducers synchronize and exchange address and data items with each side separately. Thus, transducers break end-to-end transport paths presented to higher layers into independent point-to-point segments that are routed over disconnected logical links.

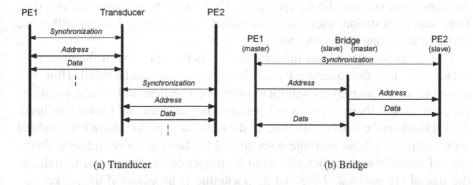

(a) Tranducer   (b) Bridge

FIGURE 3.22 Communication elements

A bridge, on the other hand, connects two busses directly at the protocol level and as such is transparent to upper layers. Bridges do not perform any buffering of data and do not participate in any higher-level protocol functions. Instead, as shown in Figure 3.22(b), end-points have to bypass the bridge to perform any necessary synchronization before the actual transfer through the bridge. A bridge always has exactly two interfaces: a master side and a slave side. As shown in Figure 3.22(b), a bridge serves transaction requests received on its slave (*PE2*) side by performing corresponding shadow transactions on its master (*PE1*) side, interleaving the master transaction into the slave transaction in the process. Therefore, a bridge has to be able to split each slave protocol transaction in two parts: listening for addresses and serving data. For a read transaction as shown here, the bridge receives a slave request and performs the corresponding master transaction before answering the slave side with the obtained data.

Note that a bridge can only connect two busses that are sufficiently timing and protocol compatible. In contrast, a transducer can connect any two (or more) busses in any master and/or slave combination. Due to the lack of buffering, synchronization and routing, however, bridges are generally cheaper and simpler to implement. In either case, the network layer inserts any necessary transducers or bridges in order to correctly and optimally connect different busses in the system.

3.5.5 TRANSPORT LAYER

The transport layer sits on top of the network layer, and together, they are responsible for transmitting messages between communication endpoints over the network of logical links and CEs. Depending on the topology, the network might have to perform buffering of data in intermediate transducers. In order to reduce the required sizes of these buffers, the transport layer splits large messages into smaller chunks (packets). Packets get transferred one at a time from station to station, such that intermediate stations only have to buffer one packet, rather than the whole message, at any time.

Part of the transport layer implementation is the selection of the maximum packet size. In order to reduce buffer sizes, packets should be small. However, since the underlying layers will incur overhead (headers and synchronization) for each packet, there is a trade-off between buffer sizes and packet overhead. Packet sizes can be fixed or variable. In the fixed case, packets have to be padded with empty data if the message does not fill the last packet completely. In the case of variable sizes, a header has to be prepended to the packet to indicate the size of the payload. Note that no packeting is necessary if the packet size is bigger than the size of the largest message.

Since intermediate buffering in the network decouples endpoints and destroys any synchronicity, the transport layer also has to potentially restore any such bidirectional end-to-end dependencies. If the application requires synchronous data transfers and if there are buffers in the path between endpoints, the transport layer will have to perform additional end-to-end synchronization by exchanging acknowledge messages at the end of each message.

Finally, in combination with the network layer, switching and routing of packets over the CEs in the network have to be implemented. Each transducer in the network has to be able to determine which end-to-end path a certain packet belongs to, either by allocating a dedicated incoming and outgoing link for each separate end-to-end path, or by adding identifying headers with endpoint addresses to each packet. Based on the association with end-to-end paths, a decision can then be made in the transducers as to where to route each incoming packet. If there is only one possible path between a pair of endpoints, routing is straightforward. In all other cases, packets can either be routed statically or dynamically using, respectively, a pre-determined, fixed route or a route that is dynamically computed during runtime based on information about the current or estimated network status.

3.5.6   LINK LAYER

The link layer provides logical links for point-to-point packet transfers between adjacent (directly connected) stations of the network. For each packet transfer it can be observed that:

(a) One station is the sender and the other station is the receiver of the data.

(b) One station is the master (which actively initiates/controls the transfer) and the other station is the slave (which passively waits and answers transfer requests).

Note that sender/receiver and master/slave designations for stations are independent of each other. Moreover, stations can have varying designations for each transfer they perform, even between the same partners. For example, station often have to switch between sending and receiving. But stations can also switch between master and slave mode. In the case of a typical master/slave bus, designations of stations as masters and/or slaves are fixed. By contrast, in node-based, network-oriented busses such as CAN, RS232 or Ethernet, a station is usually a master when actively sending and a slave when passively receiving data. Master/slave behavior is thus coupled to the direction of the data transfer.

FIGURE 3.23 Link layer

Given these observations, the link layer is responsible for implementing synchronization from slaves to masters before performing the actual data transfers through lower layers (Figure 3.23). The link layer has to implement a mechanism to notify the master that the slave is ready before the master is allowed to initiate the transfer. Note that if there are different logical sources of requests (e.g. multiple logical links from the same slave station), separate synchronization has to be implemented for each source such that the master can safely determine which request it is allowed to initiate.

If it is guaranteed that a slave will always be ready to answer requests, synchronization is not needed, as is the case for memories or memory-mapped I/O in hardware slaves (i.e., HW-internal registers mapped onto the bus, as seen in Figure 3.17(c), Section 3.5.1). A further exception is the case where full two-way synchronization is built into the low-level protocol, such as in an RS232 bus with hardware handshaking. In such cases, no explicit synchronization is necessary and the link layer is empty. In all other cases, a dedicated synchronization mechanism from slave to master becomes necessary.

In cases where two communication partners are fixed in terms of their master/slave assignments (e.g. HW/SW communication where CPUs can only be bus masters), the most efficient way of synchronizing the two partners is to include a separate, out-of-band connection between the components, purely for synchronization purposes. Examples of separate signaling mechanisms from slave to master include typical interrupts, which are, for example, supported by most standard processors.

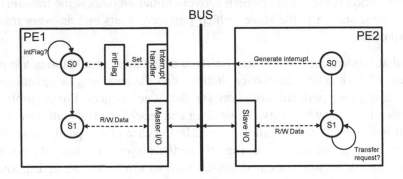

(a) Dedicated interrupts

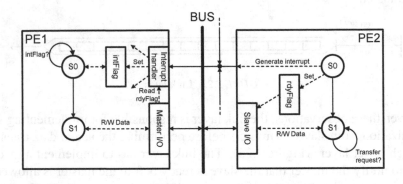

(b) Shared interrupts

FIGURE 3.24 Link layer synchronization

In the simplest, most straightforward interrupt implementation, a dedicated interrupt connection is available for and assigned to each logical link between components. In this case, two components synchronize via an event on the dedicated interrupt connection. As shown in Figure 3.24(a), on the slave side (*PE2*), the link layer implementation first sends the interrupt in state *S0* before entering state *S1* to wait (through the slave bus interface) for the data transfer to be initiated and performed by the master. On the master side (*PE1*), a dedicated interrupt detection and handling logic running concurrently to the main thread continuously listens for events on the interrupt input and sets a local *intFlag*, indicating that the slave is ready whenever an interrupt is received. The link layer in the master then always first waits (state *S0*) until the *intFlag* is set before resetting the flag and performing the actual data transfer (state *S1*) through its bus mastering interface.

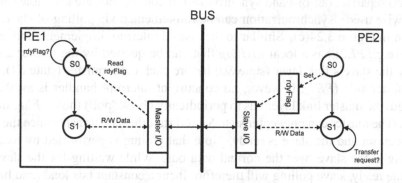

(c) Slave polling

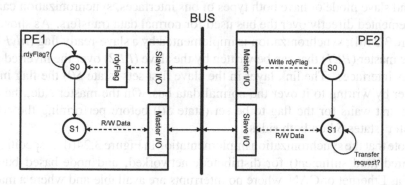

(d) Flag in master

FIGURE 3.24 Link layer synchronization (con't)

In cases where master/slave settings for communication partners are fixed but not enough interrupt connections are available, interrupts can be shared across different links. For example, Figure 3.24(b) shows a case where the same master interrupt input has multiple slave sources attached to it. Compared to an implementation with dedicated interrupts, each slave (e.g., *PE2*) has a local ready flag. When sending an interrupt event in state *S0*, a slave always sets its local *rdyFlag*. In addition, the slave maps the *rdyFlag* onto the bus such that its contents can be queried by the master. Whenever an event is received on the master side (*PE1*), the interrupt handler reads and resets the ready flags of all associated slaves to determine which slave(s) have waiting requests. Based on these results, the interrupt handler then sets the corresponding local flag(s) in the master. Thus, any enabled link layer transitions from state *S0* to *S1* are triggered as previously described for the case of dedicated interrupts.

If no separate, out-of-band synchronization connections are available (or are otherwise used), synchronization can be implemented via polling of slaves, as shown in Figure 3.24(c). Similar to the shared interrupt implementation, each slave (e.g., *PE2*) has a local *rdyFlag* that can be queried by the master and is set by the slave's link layer (state *S0*) before each data transfer (state *S1*). On the master side (*PE1*), however, no concurrent interrupt handler is available. Instead, the master link layer has to periodically check (poll) the *rdyFlag* in the slave. The master can only enter state *S1* and start the data transfer once the flag has been set and the slave is ready. Note that polling is performed by reading the flag in the slave over the normal data bus. While waiting for the slave to become ready, slave polling will therefore incur a constant bus load (and hence overhead), as determined by the chosen polling period.

Finally, in cases where communicating components can switch between master and slave mode or have both types of bus interfaces, synchronization can be implemented directly over the bus used for normal data transfers. As shown in Figure 3.24(d), synchronization is implemented by a slave-ready flag (*rdyFlag*) in the master (*PE1*) that can be written by the slave (*PE2*) over the inverted pair of bus interfaces. The link layer in the slave first sets (state *S0*) the flag in the master by writing to it over the normal data bus. On the master side, the link layer first waits for the flag to be set (state *S0*) before performing the actual transfer (states *S1* on both sides).

Note that the synchronization implementation in Figure 3.24(d) is specifically required (and sufficient) for distributed, networked, and node-based busses, such as Ethernet or CAN, where no interrupts are available and where a master can not actively receive or request data. Such synchronization via master flags can also be beneficial in cases of regular master/slave busses if communication partners already implement both types of interfaces. In contrast to an interrupt-based implementation, no specific interrupt detection and generation logic is necessary. Furthermore, in contrast to polling, which introduces a regularly triggered bus load, bus traffic overhead for synchronization is minimized.

In summary, using any of the available mechanisms to match application requirements and target architecture capabilities, the link layer implements all synchronization as it is necessary to realize logical links and packet transfers between master and slave stations in the network. As a result, the link layer separates point-to-point packet transfers into control transactions for synchronization and data streams for implementation in lower layers.

3.5.7   STREAM LAYER

The stream layer is responsible for multiplexing of different data streams going over an underlying shared medium. In order to merge multiple data sources, the stream layer has to be able to separate data packets from different streams corresponding to different logical links mapped to the same medium. Data packets are separated by assigning them addresses, where different packet streams carry different addresses. Note that packet transfers in the same stream are already separated in time and can hence share the same address. Based on a packet's address, stream layer implementations on both sides can then internally distribute the packet from the right source and to the right destination.

In the most straightforward case, each packet stream going over a shared medium/bus is assigned a unique physical bus address. Data transactions on the shared medium carry the stream's address to distinguish them from each other. Note that a PE/CE might be assigned multiple addresses if there are multiple concurrent and overlapping streams going in and out of it, either to different partners or even between the same pair of PEs/CEs.

In cases where there are not enough physical addresses available, the same physical address can be shared among multiple concurrent packet transfers going in and out of a PE/CE. At minimum, however, each pair of communicating stations on a shared bus requires a different physical address. In order to distinguish packets of different streams in case of shared physical addresses, a stream ID has to be prefixed to the packet data and transferred as part of the packet header.

All combined, resolution of addressing together with splitting of packet transfers into control and data transactions by the link layer results in unified streams of control events and data blocks being exchanged between stations at the output of the stream layer.

3.5.8   MEDIA ACCESS LAYER

The Media Access (MAC) layer is the first layer to provide an immediate abstraction of the shared underlying medium. At its canonical interface, the MAC layer represents the underlying bus as a single, shared medium over which packet data in the form of untyped blocks of bytes can be transferred. Internally, the media access layer slices the data blocks into different transactions supported by the underlying bus protocol. For example, Figure 3.25 shows the byte stream originally generated by the presentation layer (see Figure 3.19, Section 3.5.2) being split into bus word write transactions strictly two bytes at a time.

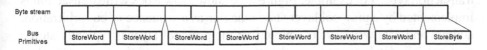

FIGURE 3.25   Media access layer

The MAC layer is thus responsible for using available protocol transactions (e.g. burst modes) in an effective and optimal manner. Together with the protocol, the media access layer also handles proper access control and locking (i.e., arbitration) to resolve and manage conflicting (overlapping) accesses to the shared medium by different streams. In general, arbitration can be performed in either a centralized or distributed fashion. In the case of a centralized scheme, additional arbiter components are part of the system busses and are inserted into the system architecture as part of the protocol layer implementation. By contrast, in a distributed arbitration, bus protocol implementations contain additional support for resolving conflicting accesses internally among themselves.

Note that while its interface to higher layers is canonical and independent of the underlying medium, the implementation of the media access layer depends to a large extent on the actual protocol characteristics such as the data widths and transaction types supported by the bus. As such, the MAC layer is the first bus-specific layer. Hence, its interface separates and translates between target-dependent aspects in lower layers and application-specific code in higher layers.

3.5.9   PROTOCOL AND PHYSICAL LAYERS

At its interface to higher layers, the protocol layer provides services to transfer groups of bits over the actual bus. The protocol layer thereby provides support in the form of services or primitives for all transaction types supported by the bus. Different transactions can thereby, for example, vary in the size of words or frames being transferred. The interface provided by the protocol layer is therefore highly dependent on the underlying bus in terms of number, types, semantics, and parameters of supported bus primitives.

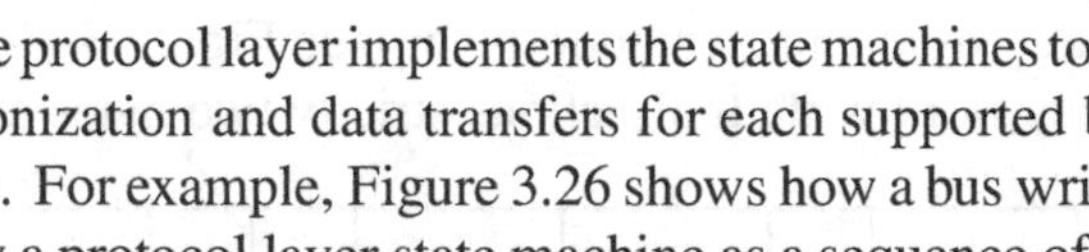

Bus Arbitration	Bus Arb. n+1	Bus Arb n+2
Addr. Cycle n-1	**AddressCycle**	Addr. Cycle n+1
Data Cycle n-2	Data Cycle n-1	**DataWriteCycle**

time →

FIGURE 3.26   Protocol layer

Internally, the protocol layer implements the state machines to perform access control, synchronization and data transfers for each supported bus primitive or transaction type. For example, Figure 3.26 shows how a bus write transaction is implemented by a protocol layer state machine as a sequence of bus arbitration, address and data write cycles. Note that the protocol layer is typically stored in some bus database and is a direct implementation of the bus protocol as described, for example, in the form of timing diagrams in the datasheet of a bus.

Finally, the physical layer is responsible for the transmission of raw bits over a physical communication channel. It defines and implements the represen-

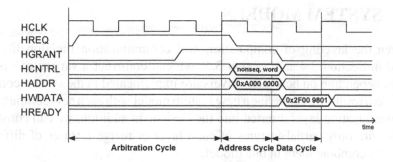

FIGURE 3.27 Physical layer

tation of bits on the physical medium. In case of a typical bus, the physical layer introduces the pins and wires of the bus and defines the corresponding voltages and bit timing. For each protocol cycle, the physical layer then drives and samples the wires correspondingly such that communication partners can exchange the selected bit values over a physical distance. For example, Figure 3.27 shows the refinement of the general protocol layer state machine from Figure 3.26 (in this case, for the example of an AMBA AHB bus [4]). For each arbitration, address or data cycle in the protocol layer, the physical layer performs the required signal transitions on each wire of the bus. In a protocol layer arbitration cycle, the physical layer raises the *HREQ* request signal and samples the *HGRANT* wire coming from the arbiter. Likewise, in the address and data protocol cycles, the physical layer is reponsible for driving and sampling the *HCNTL* control and *HADDR* or *HWDATA* address or data wires of the bus, respectively.

On the whole, protocol and physical layers implement external interfaces used to connect system components together. Hence, in the process of inserting protocol and physical layers, the wiring of component ports to bus signals is resolved, generating the final system netlist at the output of the overall communication design flow.

All together, organization of communication functionality into stacks of communication layers as described in the previous sections is the basis for developing corresponding models of system communication. Layers are separated by well-defined interfaces and each layer defines a set of features and aspects of communication behavior that need to be modeled. Similar to processor layers (Section 3.4), a model of system communication can be constructed in a layer-based fashion. With each new layer, additional communication functionality is described and specified, all the way from high-level application primitives down to bus transactions and eventually bus pins and wires.

3.6   SYSTEM MODELS

Given the layering of computation and communication functionality described in Section 3.4 and Section 3.5, we can construct a variety of system
models, depending on how many layers are implemented in the system components. As a result, we can define a new system model with each new computation
or communication layer inserted into the design. In addition, we can construct
models with only partial features of each layer or merge features of different
layers and combine them in one model.

In general, as outlined throughout Section 3.4 and in Table 3.2, communication layers are implemented inside matching computation layers depending on
the relationship and inherent dependencies of the former on services provided
by the latter. For example, the link layer becomes part of the OS layer where it
can rely on interrupt servicing mechanisms provided by the OS and the HAL.
Therefore, when constructing system models, communication layers are generally inserted either after or together with the corresponding computation layers
that hold them.

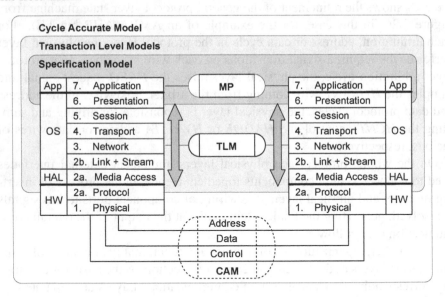

FIGURE 3.28 System models

As shown in Figure 3.28, we can generally distinguish three different classes
of models needed for different types of designers: specification models for
application programmers, Transaction-Level Models (TLMs) for system designers, and Cycle-Accurate Models (CAMs) for implementation designers.

Application designers define the system specification as system behavior
described in different MoCs, typically in some form of a hierarchical, sequen-

tial/parallel composition of processes communicating through abstract variables and message-passing channels. The specification model then provides the application designer with feedback about the match between the desired behavior and the requirements and restrictions imposed on a potential target platform in terms of needed processors and memories.

System designers subsequently define the details of the system architecture and the implementation of the application on the platform. Application processes and variables are mapped to PEs and processes on different processors communicate with each other through channels mapped onto busses. System designers require models that allow them to validate corresponding design decisions by incorporating resulting layers of functionality. As previously shown in Table 3.2, the level of communication layers in the model thereby determines the semantics of system communication required between components. Models for system designers are TLMs on an abstraction level above pins and wires. Depending on the number of layers inserted into components, a variety of TLMs can be generated. With each new layer, component interfaces are refined to communicate through services usually provided by the next lower layer. Instead, however, components in a TLM communicate through system channels that provide equivalent interfaces, services and semantics. Such TLM channels thereby describe communication between components in an abstracted form, encapsulating functionality of all layers that are not yet represented.

For example, in a TLM down to and including the network layer, individual stations in the network communicate by exchanging packets with their immediate neighbors over logical point-to-point links. On the other hand, a TLM at the interface to the protocol layer will include models of bus channels at the granularity of individual bus read/write transactions.

Finally, implementation designers define and validate the implementation of components at their microarchitecture level. Consequently, they need models that are cycle-accurate. In a CAM, final protocol and physical layers are inserted and individual bus address, data and control signals, pins and wires are observable. Furthermore, hardware and software components are further refined down to cycle-accurate RTL or instruction-set descriptions, respectively.

3.6.1 SPECIFICATION MODEL

The specification model describes the desried system functionality to be implemented at the input of the system design process. As such, a specification model is equivalent to a description of system behavior in a generic, process- and state-based MoC such as a PSM. For example, Figure 3.29 shows a simple specification model with processes *P1* and *P2* that communicate via a set of abstract, typed message-passing channels in between them.

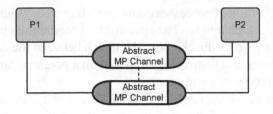

FIGURE 3.29 Specification model

The specification model can be used by application designers to prove the validity, feasibility and requirements of the application behavior. Through profiling or other analysis, the specification model can provide initial feedback about basic implementation-independent characteristics. Such specification metrics can be used both to optimize the application and to drive the system design and exploration process.

Furthermore, processes can be annotated with estimates of their execution time or energy profile on a given set of target processor candidates. As such, the specification model can include initial performance, power consumption or other quality metrics of the application running on a potential target platform, all with little to no additional overhead compared to a purely functional simulation of the application only. This provides application designers with early feedback about effects of application characteristics, such as the amount of available parallelism, on target plaform capabilities and requirements, such as the number and type of available processors.

All together, the specification model serves as the basis for developing applications both in terms of their behavior as well as considering their overall implementability at the very start of the system design and exploration process.

3.6.2   NETWORK TLM

The network TLM is the first model that reflects the overall topology of the system architecture and the final communication network. The model is a netlist of PEs, memories and CEs connected by abstract, universal bus channels (Figure 3.30). In the network TLM, processes, variables and complex channels of the system behavior are mapped onto PEs, memories and busses of the system platform. As a result, the network TLM contains application layers for both computation and communication. Processes are grouped under application layers of processors according to the given mapping. Application layers of communication resolve synchronization, storage and complex channels between processes down to the level of canonical primitives supported by lower layers.

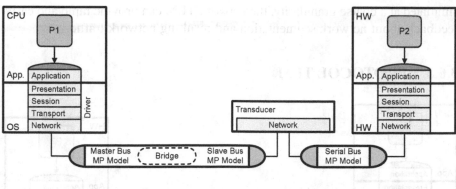

FIGURE 3.30 Network TLM

Application layers are then integrated into a network-level model of the computation and communication platform. OS layers of all software processors are introduced. On the HW side, OS layers do not exist, and initial versions of the hardware layer are inserted instead. Finally, communication functionality is implemented down to and including the network layer, inserting presentation, session, transport and network layers into the components. In the process, models of transducers that break the network into segments and route data are added. Note that on the CPUs, inserted layers are integrated with the OS model to become the drivers for implementation of reliable end-to-end communication with other processors.

Component models in the network TLM communicate via point-to-point logical links implemented over abstract, universal bus channels that each represent one segment of the network. A universal bus channel provides services for multiple simultaneous packet transfers, blockwise memory accesses and event notifications. Using the channel, components within a segment can communicate by performing transfers with message-passing semantics. At its interface, the bus channel allows components to use an arbitrary number of logical addresses in order to establish corresponding logical links and memory interfaces with their neighbors.

Internally, the code in the universal channel is independent of the implementation of the bus. Instead, it provides the functionality of link, stream, media access, protocol and physical layers in an abstract manner at a granularity of blocks of data or events. The bus channel can include estimated timing for each individual transfer based on the bus bandwidth in relation to the size of the data block. Since it is operating at a level of whole blocks, however, the universal bus channel cannot take into account effects of interleaving and scheduling of overlapping bus transfers as regulated in reality by arbitration on any underlying shared medium. Hence, the overall timing estimation within each network segment and across the whole system is approximate. Yet, since bus traffic is simulated at a coarse granularity, the network TLM can provide rapid and early feedback about network segmentation and resulting network traffic.

3.6.3 PROTOCOL TLM

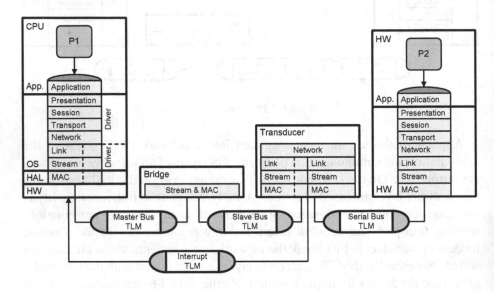

FIGURE 3.31 Protocol TLM

The protocol TLM provides an abstract, virtual prototype of the system that more accurately reflects the complete system computation and communication structure. The protocol TLM implements all application-specific functionality down to and including the HAL and MAC layers. It only abstracts away target-specific details of pre-defined hardware including protocol and physical layer bus interfaces. As shown in Figure 3.31, the protocol TLM adds link, stream and MAC layer implementations in each PE, memory, and CE. In the process, actual models of bridge CEs are introduced.

More specifically, on the CPUs, all drivers are generated on top of the MAC instance that, together with the HAL, is taken out of the PE database. Generated drivers include interrupt handlers and interrupt tasks integrated into the OS. As shown previously in Figure 3.14(a), Section 3.4.4, the CPU in the protocol TLM can include a transaction-level model of the hardware layer that accurately describes external interrupt scheduling and processor suspension.

Components in the protocol TLM communicate over bus channels that represent and abstract the actual physical medium of each bus segment. Bus TLM channels provide timed yet abstract implementations of all supported bus protocol transactions, such as byte, word, double-word or burst reads and writes. As such, exact semantics and interfaces of bus TLM channels are specific to a certain bus being modeled. Bus protocol channels can thereby model the known timing and effects of arbitration accurately at the level of individual transactions. In addition to bus channels, the TLM contains interrupt channels that abstract the interrupt wires and interrupt detection and generation into event transfers used by the link layer for implementation of any necessary synchronization.

Since modeling of physical wires and signals imposes a heavy overhead in event-driven simulation, the protocol TLM executes significantly faster than a corresponding BCAM (see Section 3.6.4 below). On the other hand, since all user-defined implementation detail is represented and all computation and communication parameters are known, the protocol TLM can provide accurate results about all design decisions to the implementation designer. Hence, the protocol TLM simulation provides fast and detailed measurable feedback about the effects of each of the layer implementations in terms of bus traffic, interrupt handling overhead or processor load.

3.6.4 BUS CYCLE-ACCURATE MODEL (BCAM)

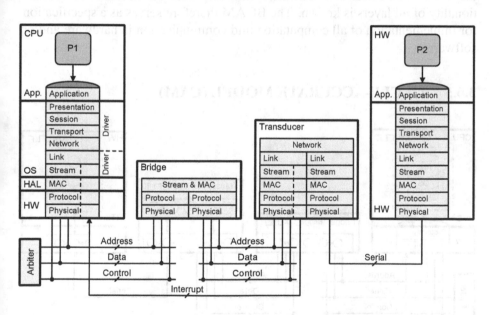

FIGURE 3.32 Bus Cycle-Accurate Model (BCAM)

The Bus Cycle-Accurate Model (BCAM) implements all layers of computation and communication functionality in all PEs, memories and CEs of the system (Figure 3.32). Compared to the TLM, the BCAM includes additional implementations of protocol and optionally physical layers. As a result, the BCAM is structurally accurate and its components communicate via cycle-accurate descriptions of system busses and interrupt lines. As part of the bus structure, the BCAM can include additional bus components such as multiplexers, repeaters or arbiters. Note that depending on whether the physical layer is included, a BCAM may or may not be pin-accurate. In the pin-accurate version, the BCAM describes the complete system netlist as a set of blocks connected by signals. Otherwise, physical layers are not included and components communicate through transactions at the level of individual, abstracted address, data and control cycles.

In contrast to other models, actual transitions and events on the busses can be observed in the BCAM on the basis of individual bus cycles. Hence, the BCAM provides limited additional accuracy to model cycle-by-cycle behavior needed in cases of advanced busses with support for split transactions or for preemption of transfers by high-priority masters. Note, however, that the BCAM requires significant additional overhead for simulation of individual events in every cycle.

The BCAM is the basis for cycle-accurate implementation of components in the backend hardware and software synthesis processes. In the BCAM, functionality of all layers is known. The BCAM therefore serves as a specification for implementation of all computation and communication in hardware and/or software.

3.6.5   CYCLE-ACCURATE MODEL (CAM)

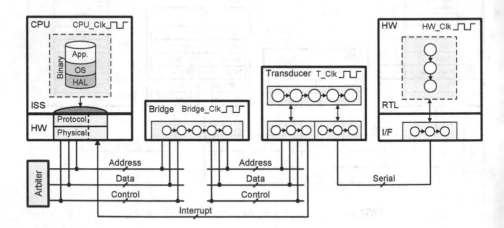

FIGURE 3.33   Cycle-Accurate Model (CAM)

The Cycle-Accurate Model (CAM) is the result of the hardware and software synthesis process in the backend (see Chapter 5 and Chapter 6) and hence the final result of the system design process. As shown in Figure 3.33, models of all components in the system architecture are refined down to their cycle-accurate representations. On the software side, behavioral CPU models in the BCAM are replaced with ISS models that execute final binaries with target-specific object code for application, OS and HAL, including all drivers and interrupt handlers. ISS models are encapsulated in a wrapper that integrates the software simulator into the system by relaying all external bus and interrupt accesses through corresponding instances of protocol and physical layer hardware models.

On the hardware side, protocol and physical layers are replaced with cycle-accurate FSMs that drive and sample bus wires to implement all bus protocol transactions. Upper layers, up to and including the application itself, on the other hand, are synthesized into corresponding FSMD models of their hardware implementation at the RT level. For external bus accesses, main FSMDs communicate with bus interface FSMs through a register and signal interface.

Finally, bridge and transducer CEs are similarly synthesized into cycle-accurate state machine models. Bridges are realized as a product state machine that translates between two protocols at the level of common bus cycles. Transducers, on the other hand, contain FSMs for each bus interface and a main FSMD that implements queues and buffers for independent synchronization and forwarding of data transfers between interfaces.

All in all, the CAM provides implementation designers with an environment for validation of component realizations at the RTL architecture level. In addition, the CAM allows for a cycle-accurate simulation of the complete system as a final signoff before further logic and physical synthesis.

3.7 SUMMARY

In this chapter we presented concepts and techniques for modeling of both system computation and communication at various levels of detail. Starting with the system behavior described in terms of different MoCs and captured in a system design language, system functionality is gradually refined down to an implementation by inserting new layers of implementation detail with each design step. In the process, processor and communication layers are combined into complete system models at different levels of abstraction. Each abstraction level thus includes different design decisions and provides validation of different design metrics. All in all, the flow of well-defined system models provides the basis for an automated design process with support for synthesis and verification in each step.

As typical modeling results in Figure 3.34 show, we can generally observe a trade-off between simulation speed and accuracy of results when modeling systems at varying levels of abstraction. The graphs here show both the simulation times (Figure 3.34(a)) and the average error in simulated frame delays (Fig-

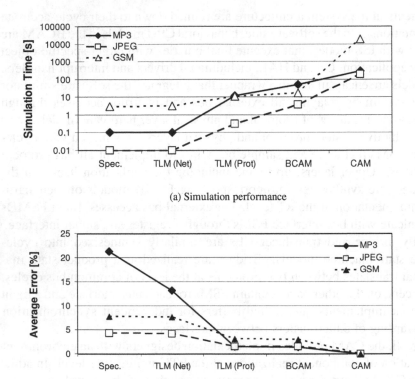

(a) Simulation performance

(b) Accuracy

FIGURE 3.34 Modeling results

ure 3.34(b)) obtained during validation of several realistic, industry-strength
design examples, such as an MP3 decoder, a JPEG encoder and a GSM voice
encoder/decoder for mobile phone applications. In all cases, results confirm
that simulation times generally increase exponentially while accuracy increases
linearly, as more implementation detail is added with lower levels of abstraction.

A traditional CAM running cycle-accurate software ISSs next to hardware
RTL models is 100% cycle-accurate but slow. Note that ISS approaches exist
that are significantly faster but provide only approximated or no timing [88,
146]. On the other hand, a purely functional specification simulates at native
speeds but with no accuracy. In between, TLMs at various levels support
trade-offs between speed and accuracy. Specifically, a protocol TLM can be
as accurate as a BCAM for systems that do not utilize busses with complex
arbitration schemes such as preemption or split transactions. A network TLM
can provide the same accurate feedback only if there is no arbitration and busses

each have a single master only. In both cases, TLMs simulate at much higher speeds compared to a BCAM or CAM.

As such, TLMs are ideal candidates for virtual prototyping of the design and corresponding rapid, early design space exploration. Complementing traditional CAM simulations, a variety of TLMs at different levels of detail enable gradual pruning of the design space as design progresses from a specification down to its final implementation. In addition, note that simulations with an arbitrary combination of cycle-accurate or behavioral component models are possible at both the protocol TLM and BCAM/CAM level. Such mixed-level co-simulations allow for further speed and accuracy trade-offs by validating structural component implementations embedded in a fast behavioral simulation of the system environment.

Chapter 4

SYSTEM SYNTHESIS

Synthesis is one of the key automation techniques for improving productivity and developing efficient implementations from a design specification. Synthesis refers to the creation of a detailed model or blueprint of the design from an abstract specification, typically a software model of the design. Synthesis takes different forms during different stages of the design process. In hardware system design, several synthesis steps automate various parts of the design process. For instance, physical synthesis automates the placement of transistors and the routing of interconnects from a gate level description, which itself has been created by logic synthesis from a register transfer level (RTL) model. The same principle of translation from higher level model to a lower level model applies to system synthesis.

Embedded system designs consist of both software and hardware. Although the synthesis principle stays the same, the number of design decisions increases substantially and their impact becomes less predictable. This makes the system synthesis process more complicated than it is for hardware synthesis alone. To meet this complexity, designers can adopt a model based approach, in which they create unified models of both the software and hardware in the design. Such system level models do not distinguish between software and hardware. Instead, they capture the functionality of the application and the structure of the computational platform. Typically, these models are executable, so they may be validated by simulation.

In Chapter 3, we looked at different modeling abstractions for embedded systems. We saw that the system platform consists of processing elements such as CPUs, DSPs, and custom HW IPs, connected via a network of buses, links, routers, and bridges. The application is mapped on this platform. We also defined formal rules and semantics for models at various abstraction levels including specification, TLM and CAM. In a model based design approach, we

D.D. Gajski et al., *Embedded System Design: Modeling, Synthesis and Verification*,
DOI: 10.1007/978-1-4419-0504-8_4,
© Springer Science + Business Media, LLC 2009

define system synthesis as the generation of a TLM from a specification model. Therefore, formal model semantics are essential for identifying synthesis requirements and developing synthesis procedures.

In this chapter we will present methods and techniques for system synthesis. We will first examine trends in system design, including past board-based methods and state of the art virtual platform based design. We will then present a model based approach that enables automatic system synthesis, and the three techniques central to this approach: model generation, mapping generation and platform generation. These techniques can be used to automate the synthesis of embedded systems from a high level specification model.

4.1 SYSTEM DESIGN TRENDS

Traditional system design starts with a platform definition as shown in Figure 4.1. The platform architect defines the type of processors and their communication architecture at a high level, taking into account the application characteristics. For example, in multimedia codec designs, digital signal processors (DSPs) are typically used. For control intensive applications, embedded processors may be used. The number of processors depends on the number of independent tasks or available parallelism in the application. Without an evaluation model, the platform architect depends on his or her experience and the application profile to make these decisions. Very often though, the platform may be chosen based on legacy considerations. When the product is updated for the next generation, new components may be added to the platform for the additional product features.

Once the platform is defined, it is passed on to the hardware engineers who implement the platform and deliver the board. The hardware engineers develop HDL models of any custom blocks and configure the processors and other IPs. Once the hardware becomes available, the software developers can develop the system software.

Traditional system design requires the development of boards and board support packages (BSPs) before the application's SW development could begin. Typically, designers define the platform architecture and the mapping of appli-

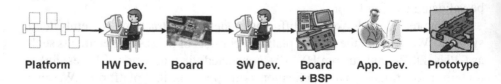

| Platform | HW Dev. | Board | SW Dev. | Board + BSP | App. Dev. | Prototype |

FIGURE 4.1 A traditional board-based system design process.

cation functions to different components of the architecture. Then, HW and system SW are developed to produce the board and BSPs. Finally, application SW is created using the BSPs and downloaded to the board to produce the system prototype. Both the application SW and the HW/BSP development take several months, which means that the prototype is not ready until more than a year after specification.

The single most common reason for delays in traditional board-based design methodology is its sequential development of hardware and software. That the software depends on hardware availability also leads to verification issues. For instance, if a bug is found during software development, the software engineers typically place the blame on buggy hardware. Conversely, the hardware engineers contend that the bug is due to the software team's poor understanding or improper use of the hardware. Such problems may lead to unnecessary interactions between the design teams, ad-hoc decisions and wasted time, resulting in a product delivery that may be seriously delayed. Virtual platform based design methodology which in Figure 4.2 breaks down, is one way to avoid the delays.

A virtual platform based design methodology begins with the idea that a model of the hardware platform may be used for software development, rather than a prototyping board. This model is called a virtual platform (VP). The benefit of a VP is that it takes less effort and less time to develop than a prototyping board because abstract models of the components may be used to create the VP. The internal micro-architecture of the processing elements may not be defined in a VP. However, the VP must provide a programmable model of all software processors and functional models of all custom hardware components. Typically, these models are at a level of abstraction higher than cycle accurate, though they still provide visibility into the processor registers and bus transactions for debugging and run-time analysis of the embedded software.

Virtual platforms are usually implemented using C/C++ models of the processors. The hardware peripherals are modeled as remote function calls, which provides the simulation speed that is crucial for rapid embedded software development. With the advent of system level design languages, such as SystemC,

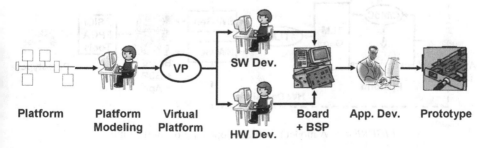

| Platform | Platform Modeling | Virtual Platform | SW Dev. HW Dev. | Board + BSP | App. Dev. | Prototype |

FIGURE 4.2 A virtual platform based development environment.

VPs for multicore architectures can be modeled. Since the embedded SW can then be developed on the VP, it can be done in parallel with the hardware. Thus, the VP serves as a common golden reference for the development of both the embedded software and the hardware modules. Since the software is available earlier, it may even be used for realistic debugging of the hardware modules. The result is that the board and its support package (BSP) as shown in Figure 4.2. This methodology shortens overall prototyping time. However, development of VPs must still be done by engineers with expertise in SystemC and platform modeling. Another disadvantage is that any changes to the platform must be implemented manually in the VP. During product upgrades, the embedded software may need to be rewritten and tested for the new platform. This is a significant undertaking, given the move to multicore platforms and the rise in the software content of embedded systems.

These shortcomings have led to a need for system level technologies for design of embedded systems. A model-based design methodology, as shown in Figure 4.3, can succeed beyond what is possible with a virtual platform based methodology. The key model for this new approach is the TLM [31]. If the semantics of the TLM are well defined, it is possible to automatically generate the TLM from a high level, graphical description of the platform and application. Unlike the previous methodologies, in a model based design methodology, application development and porting is not postponed to the end of the design cycle. Instead, it is inputted in the design process, driving the selection of the platform as well as the generation of the embedded software and hardware.

Using this model based method, the application may be defined using C/C++ models. However, other languages and models of computation such as UML and Stateflow may be used for different application domains. The platform may either be defined by the application developer of derived automatically from the application profile. Next, a mapping from the application to the platform is cre-

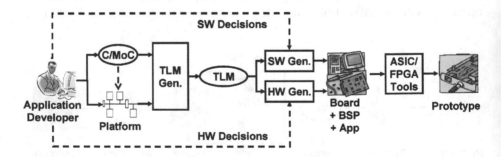

FIGURE 4.3 A model based development flow of the future.

ated. Based on this mapping, an executable TLM of the system is automatically generated.

With well defined TLM synthesis semantics and software/hardware design decisions, designers will be able to automatically generate the embedded software and RTL hardware description from the TLMs. Furthermore, modification of the system architecture or application is possible even late in the design process because new TLM, HW and system SW will be generated automatically.

4.2 TLM BASED DESIGN

TLMs are central to the model based design methodology shown in Figure 4.4. The input to the design process is the system specification model. The specification consists of an application model mapped to a software/hardware platform. The application model is a purely functional model without any implementation details; typically, it is an executable model that may be compiled natively on the host machine for early functional validation. Various models of computation may be used to define the application, including stateflow for control-intensive reactive systems, dataflow for multimedia applications, concurrent sequential processes (CSP) [90], Petri-nets [142], and Message Passing (MP) models [80] for concurrent and distributed applications.

The platform description, on the other hand, is not executable. Instead, it defines all the component instances and their connectivity using a declarative language. The components include software and hardware processors, buses, bridges, and memories. The software platform includes real-time operating sys-

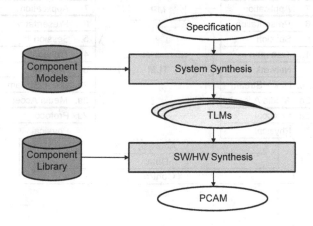

FIGURE 4.4 TLM based design flow.

tems (RTOS) and drivers for common peripherals and networking. A mapping is defined from the computational elements of the application to the processors in the platform. If the application is concurrent, then the communication elements are mapped to buses or routes in the platform. Several detailed decisions need to be made while defining this mapping, as we shall see later in this chapter.

The system synthesis tool generates the model in correspondence with the application and its mapping on the platform. In some cases, if the mapping or platform is not defined, the synthesis tool may be required to generate the appropriate mapping or platform, based on the application. In almost all design scenarios, it is important to maintain a database of components that may be used to define the platform. Such a database contains the data models of the components. In other words, the characterization of intellectual property (IP) blocks may be used to assemble the platform. The various characteristics of the components are modeled here, including their configuration settings, metrics and services.

The output of system level synthesis is the TLM. There are various types of TLMs possible, depending upon the design methodology and the availability of component models. In Chapter 3, we discussed several TLMs at different abstraction levels. The ideal TLM is one that provides a reasonable balance between the speed of execution and the accuracy of the design performance estimation.

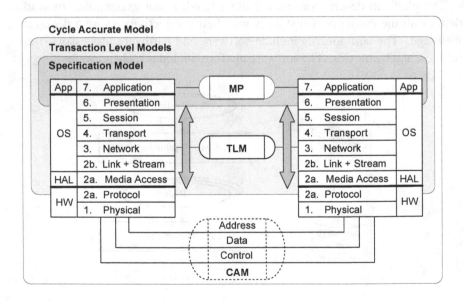

FIGURE 4.5 Modeling layers for TLM.

Since we can create several possible abstractions for TLMs, we need a way to categorize them and to define their semantics according to well known communication design concepts. The OSI standard [98] is one such categorization criteria; it proposes seven well-defined different layers for networks. Figure 4.5 shows the positioning of the specification model, the TLM, and the Cycle-Accurate Model (CAM) with respect to these OSI defined network layers.

If we apply the OSI standard, a data communication between two processes assigned to two processing elements in a platform must satisfy the requirements of these seven network layers. The specification model corresponds to the application layer or abstraction, in which there is no notion of a platform. Two processes in the Specification model communicate through abstract channels using send and receive functions which transfer data from one location to another. These abstract channels do not need any implementation details, such as routing paths or addressing. Each channel instance represents a unique point-to-point transaction media for the processes in the application. Furthermore, the abstract channels may also have abstract types, including base types, arrays or structures. In other words, no aspect of the communication is modeled other than the abstract data transfer at the specification level.

At the other end of the OSI spectrum, the CAM communication model belongs to the physical abstraction layer. Each pin and signal in the system is modeled explicitly. Data transactions can be observed as the wiggling of bits on signals that connect two processing elements or on a shared bus. Each transaction is addressed if a shared bus is used. Furthermore, cycle accuracy is observed in the bus transactions. The bus protocol defines the control, data and address signals, as well as the sequence of ordered events in the signals. Therefore, the CAM accurately represents the system on a detailed signal level and takes into account signal changes in each clock cycle. It is used to synthesize the system with standard design automation and software tools.

A TLM can be made more or less accurate by including more or fewer network layers, A TLM may model only a subset of functions in a particular layer, depending on the metrics to be estimated. Communication modeling is one of the most important TLM features. Communication can be modeled at several different abstraction levels by selecting certain features and not others. For the purpose of this chapter, we have chosen the TLM that models communication at the network layer. Therefore, all decisions involving packeting and routing are implemented in the TLM. However, lower level decisions, such as synchronization and addressing, are abstracted. The transactions take place at the packet level, in which each packet is a byte array of given size. Each bus, link, router, and bridge in the platform is also modeled explicitly.

With well defined semantics for the specification, TLM, and CAM, it is possible to automatically generate models and synthesize the embedded system. We must establish the synthesis semantics before we can define the synthesis

requirements and algorithms. In the following sections, we will discuss methods
and algorithms for system synthesis based on different design scenarios.

4.3   AUTOMATIC TLM GENERATION

The typical platform-based system level design process starts with the definition of a platform of processing elements connected to the communication
architecture. The platform is usually chosen to optimize the execution of a
given application. The application tasks are mapped to processing elements
and the abstract channels are mapped to buses or routes in the communication
architecture. A model of the application, mapped to the platform is then created to evaluate the performance. This model estimates various metrics of the
design such as delay, power consumption, and reliability. Once these metrics
are obtained, the designer evaluates if they meet the constraints defined in the
specification. If the constraints are not satisfied, the designer optimizes the
application, platform or mapping. The model is then regenerated to re-check
for constraint satisfaction.

This synthesis flow, as shown in Figure 4.6, must be supported by various
languages, automation tools, libraries, as well as a design environment. In previous chapters, we discussed the languages and models of computation needed
to specify application models. However, these languages and representations
are not suitable to define the system platform. Therefore, we need a language,
preferably graphical, to input the platform netlist. Since the platform components require configuration, we must define properties and parameters for each
component. The specification environment should also allow the the mapping
from the application to the platform to be defined. Therefore, the application
model must be expressed in a form such that a mapping of application objects
may be made to the objects in the platform definition. For example, functions,

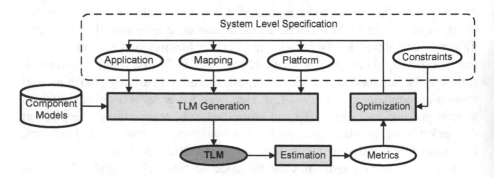

FIGURE 4.6 System synthesis flow with given platform and mapping.

tasks or processes in the application model may be mapped to software or hardware processors in the platform. The specification environment should allow the user to specify such a mapping.

Once the platform components are selected and the mapping is defined, the TLM of the design is generated. Data models of the platform components are needed in order to create a TLM. These data models have different parameters for different components in the platform. Furthermore, the parameters of the data models are also dependent on the type of metrics that we want to evaluate in the TLM. For instance, if we are interested in evaluating the execution performance of the system, we need to provide the delays associated with each operation of the processors and buses. However, if we want to estimate power consumption in the TLM, we must provide power dissipation parameters in the data model. At the very least, some basic information concerning the structural compatibility of the components must be described in the data model. For instance, the data model may specify that an ARM processor may only connect to an AMBA bus [4]. Such information in the data model will allow the designer to perform sanity checks while creating the platform. As a result, the designer will prevented from, for example, connecting an ARM processor to any arbitrary bus in the platform. Similarly, other compatibility parameters, such as supported operating systems or maximum number of threads, may also be defined in the component model.

The optimization phase of the system design process, shown in Figure 4.6, requires the definition of design constraints. Such constraints are extremely important in embedded systems, int contrast to traditional PC based systems. Almost all embedded systems must be designed to operate under constraints. Typically, there exist real-time constraints on the design, which we can classify as either hard or soft. For example, in automotive systems, the brakes must be applied on the wheels within a strict time delay after the pedal is pressed. This is a case of a hard real time constraint. On the other hand, during video or audio decoding, the delay between the decoding of two frames must meet some specified deadline. However, if the frame is not decoded in time, the quality of output may suffer, but it is not a life-threatening situation and may be tolerated. Such constraints are soft-real time constraints. For portable devices, similar constraints may exist for other metrics, such as power and energy consumption. In all cases, the constraint may be defined as a bound on a given metric or set of metrics.

In a model based methodology, a timed TLM of the design is used to check if it satisfies the given performance constraints. However, manual development of a timed TLM is difficult and time consuming. Therefore, we need methods to automatically generated TLMs. In this section, we will discuss automatic TLM generation from platform definition and the mapping of application model to the platform, and TLM generation. We will go into the details of communication and computation timing estimation to provide a sample of metric estimation techniques. We then will discuss the semantics of TLMs in a popular system level design language, called SystemC [81]. The SystemC TLM semantics allow automatic generation of TLMs from the definition of application to platform mapping.

4.3.1 APPLICATION MODELING

The design process starts with a given application. Typically, the application is specified using an executable model written in a well defined model of computation, such as those described in Chapter 3. Figure 4.7 shows a typical application model specified in the program state machine (PSM) model of computation. The PSM model gives the application developer the ability to specify concurrency, hierarchy, and abstract communication. It also allows the developer to create different flows of execution at different hierarchy levels. These flows of execution could involve sequential processing as seen in the relationship between *P3* and *P4* processes, or concurrent execution as shown in the relationship between *P5* and processes *P3* or *P4*. Finally, a PSM model may include a state machine execution flow as seen at the top level of the hierarchy.

The processes at the leaf level are symbolic representations of functions that may be specified using common programming languages such as C and C++. Even legacy code (usually available in C) can be inserted in leaf-level processes. Processes use channels for communication amongst themselves; the channels, therefore, capture data dependence between the behaviors. Communication channels enable an object oriented method of composing and connecting processes because they clearly separate computation from communication. This is a useful separation because the design of processors is orthogonal to the design of communication architecture. As a result of the separation, processes can be the input to synthesis of the processors and channels can be the input to synthesis of communication architecture. Furthermore, computation and

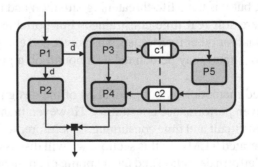

FIGURE 4.7 A simple application expressed in PSM model of computation.

communication can be debugged, modified or optimized independently of each other. The clear separation of computation and communication, therefore, simplifies model based design.

4.3.2 PLATFORM DEFINITION

Once we complete application modeling, we must implement it on a given platform. The platform is the set of software and hardware services that are provided to the application. In order to better understand the relationship of the application to the platform, let us take the example of a simple software processor. The application model is the code, written in some high level language, which must be executed on the processor. The processor definition includes an instruction set, which may be thought of as a set of services provided by the processor micro-architecture. A compiler transforms the application model into a sequence of instructions (services) so that it may be executed on the processor. The concept of an embedded system platform is similar in principle, though it is more complex than a single processor.

Figure 4.8 shows a typical embedded platform consisting of multiple CPUs (*CPU1* and *CPU2*), a hardware accelerator (*HW IP*), and memory (*Mem*). The communication architecture of the platform consists of two buses (*Bus1* and *Bus2*) connected by an interface component. The platform is usually composed from a set of components and a set of connections selected from the library or defined by the user. In general, these components can be embedded processors, memories, custom hardware units, or third party IPs. The communication architecture for the proposed platforms is also very flexible. The designer may use system level interconnects such as shared buses (with centralized or distributed

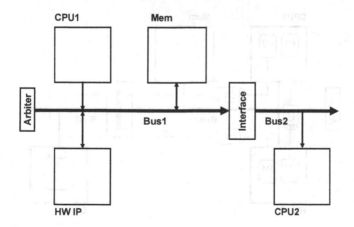

FIGURE 4.8 A multicore platform specification.

arbitration), bridges, serial links or network on chip. The platform architecture can be completely or partially defined and more components and connections can be added at a later design stage for the optimization of a particular metric. Therefore, in a model based design methodology, the platform can be easily updated for generating a new TLM.

4.3.3   APPLICATION TO PLATFORM MAPPING

So far we discussed the application modeling and platform definition steps for TLM generation. However, before the TLM can be generated, the application must be mapped to the platform. Figure 4.9 shows the mapping of the application model (in Figure 4.7) to the platform (in Figure 4.8). Note that the application has been transformed from a PSM computation model in Section 4.3.1 to a concurrent set of communicating processes. This transformation step is necessary because it allows a well defined mapping of objects in the application model to the objects in the platform.

The model of computation used to specify the application, for mapping, is a subset of the PSM model, in which the hierarchy is restricted to a single level of concurrent processes. This restriction is practical because the computation platform is a set of independent processing elements that may execute in parallel. Therefore, a one-to-one or many-to-one mapping of application processes to processing elements is possible without concern for execution flow dependence between the processes. Similarly, the communication in the PSM may be transformed. While the data communication channels in the PSM are preserved, new channels may be added to represent synchronization. An example of the synchronization channel is *C3*, which is added between processes *P1* and *P2*.

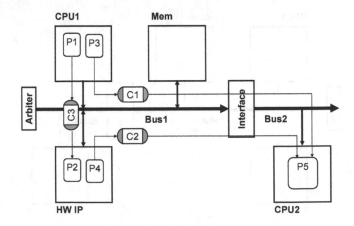

FIGURE 4.9 Mapping from application model to platform.

Note that the PSM application model in Figure 4.7 defined control dependence from *P1* to *P2*. In the transformed application model, *P1* and *P2* are expressed as concurrent processes. However, channel *C3* introduces control dependence from *P1* to *P2*, in order to preserve the original execution order in the PSM model.

This mapping has well defined rules. Processes map to processing elements (PEs) such as CPUs, HW components, and IPs. Channels between processes are mapped to routes consisting of buses, bridges, and interfaces. For the channel to be implemented, a valid route must exist between the PEs hosting the respective communicating processes. A set of possible routes for each application channel can be easily generated by analyzing the platform. The designer can then select the route for each channel and perform the mapping.

Besides the general mapping principles described above, there may be other application and platform dependent restrictions in mapping. For instance, the processor may define a maximum number of processes that may be mapped to it, as in the case of hard IP, in which the mapping is implicit because the IP model defines the processes executing on it. Likewise, the address space of the bus may restrict the number of channels mapped to it. For example, a dedicated serial link may not allow more than a single channel to be mapped to it.

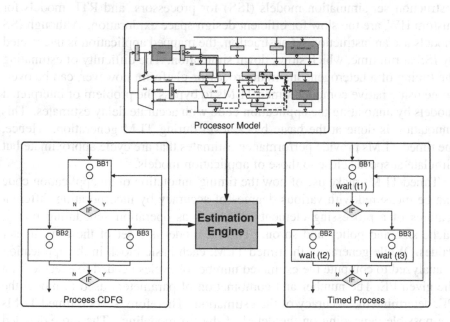

FIGURE 4.10 Computation timing estimation.

4.3.4   TLM BASED PERFORMANCE ESTIMATION

As mentioned earlier, various metrics may be evaluated by simulating the TLM. These metrics help the designer evaluate design decisions such as platform component selection and application to platform mapping. One important criteria for selecting the mapping is the performance of the system. Timing annotation in the TLM can provide us with delay estimates for the system. The timing must be annotated for both computation operations and the communication functions in the TLM.

COMPUTATION ESTIMATION

Timing estimation is not easy to perform because heterogeneous multiprocessor platforms are increasingly being used in system design to deal with the growing complexity and performance demands of modern applications. Each component model may be written in a different language or at a different abstraction level so it is often not practical, or even feasible, to compose them. Cycle accurate component models, for example, do provide accuracy but may not be available for the whole platform. Often, as in the case of legacy hardware, a C model of the component is impractical to build. Furthermore, cycle accurate instruction set simulation models (ISS) for processors, and RTL models for custom HW, are too slow for efficient design space exploration. Although ISS models use an instruction set abstraction, the mapped application is interpreted by ISS at run time, which slows down simulation. The difficulty of estimating the timing of a heterogeneous multiprocessor platform, however, can be overcome using native compiled timed TLMs to bypass the problem of interpreted models by annotating the application code with accurate delay estimates. This annotation is done at the basic block level during TLM generation. Hence, the timed TLMs provide performance estimates that are cycle approximate but simulate at speeds close to those of application models.

Timed TLMs make use of how the timing annotation of an application code can be measured with various degrees of accuracy by incorporating different features of a processing element (PE) such as operation scheduling policy, cache size and policy and so on. The PE model is a set of these parameter values. While generating the timed TLM, each basic block in the application is analyzed to compute the estimated number of cycles needed to execute it on the given PE. The number and combination of parameters used to model the PE, determine the accuracy of the estimation. Therefore, several timed TLMs are possible depending on the detail of the PE modeling. The more detailed the PE model, the longer is the time to compute execution delay. A tradeoff must be made to achieve the optimal abstraction for the PE model. The most important parameters for PE modeling are operation scheduling policy, data path structure, memory delay and branch delay.

Computation timing may be estimated and annotated automatically during TLM generation [93]. The timing annotation process is shown in Figure 4.10. Automatically generating timed TLMs allows the designer to estimate the co-mutation delays early in the design cycles. Timing annotation consists of adding timing information to the application códe based on its mapping to a given pro-cessing element. The application code in each process is converted into a CDFG representation as shown in Figure 4.10. Then, a retargetable PE model is used to analyze the execution of each basic block of process code on the given PE. This analysis provides an estimated delay for each basic block in the process. The basic blocks are then annotated with the estimated delay to produce a timed process model. This compiled estimation technique can be applied to any appli-cation code mapped to any type of processing element for which a data model exists. Therefore, TLM estimation is fast, retargettable and usually provides more accuracy than even instruction set simulation models.

COMMUNICATION ESTIMATION

Communication delay is the other critical part of the system performance. Com-munication delay becomes very relevant if the platform has a complex bus or network architecture. A high degree of data dependence between processes also contributes to the communication delay. Very often, designers are interested in end to end communication delays between processes. In other words, the designer wants to know how long it takes to send data from the sender PE to the receiver PE over the given communication architecture. If the sender PE and the receiver PE are not connected directly, the transaction may take place over a route consisting of several buses and intermediate buffers. Therefore, the end to end delay is effectively the sum of the delays on each bus segment. There-fore, it is important for us to model the delay in each bus segment accurately, in order to be able to reliably estimate the end to end delays. The principle of retarget-able estimation, as discussed above, may also be used for bus delay modeling.

Figure 4.11 shows the principle of bus delay modeling at the transaction level. Note that our notion of buses includes shared buses, network links, crossbars and serial buses. At the transaction level, we are not interested in the implementation or interconnect topology. Rather, we are interested in the types of services that the bus provides for communication. We categorize these services into synchronization, arbitration and data transfer. Arbitration may further be subdivided into two events: acquiring the bus (Get _Bus) and releasing the bus (Release_Bus). The order of actions for a given transaction is fixed as shown by the channel send function in Figure 4.11. The two communicating processes first synchronize, then the master PE attempts to reserve the bus by calling the arbitration function Get _Bus. Once the bus is acquired, the data transfer is made. Finally, the master PE releases the bus to be used by other contending PEs. As we can use the PE model for computation delay estimation, we can use a bus protocol model for communication delay estimation. We will now look into the various delays that occur during a bus transaction, and describe ways of computing them based on TLM modeling techniques and bus protocol model.

SYNCHRONIZATION MODELING

Synchronization is required for two processes to exchange data reliably through a channel. The sender process must wait until the receiver process is ready to receive. Similarly, the receiver process must not read the data until the sender has written it. This is one of the most common forms of blocking synchronization and is known as double handshake or rendezvous synchronization. There are several ways of implementing double handshake synchronization in hardware and software. However, at the transaction level, we may use higher level prim-

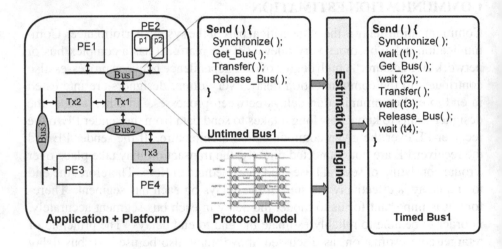

FIGURE 4.11 Communication timing estimation.

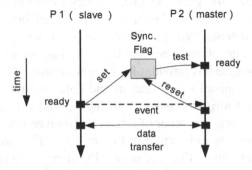

FIGURE 4.12 Synchronization Modeling with Flags and Events.

itives such as flags and events to model the synchronization. Such modeling constructs are sufficient to estimate the approximate delays for cycle accurate synchronization.

Synchronization in the bus channel may be modeled with a set of flags and events. Each flag-event pair corresponds to a unique application level channel that is routed through the given bus segment. The flags and events may be stored in a synchronization table in the bus channel and are indexed by the channel ids. The mechanism for double handshake synchronization is illustrated in Figure 4.11. Synchronization between two processes takes place by one process setting the flag and the other process checking and resetting the flag. Once the flag has been reset, the transacting processes are said to be synchronized.

We refer to the component that sets the synchronization flag as the *slave*, while the component that tests and resets the flag is the *master*. Typically, in computer system design software processors act as masters while hardware peripherals act as slaves on the bus. We will refer to the process mapped to the slave PE as the slave process, while the process mapped to the master PE will be referred to as the master process. Note that for any transaction, one process should be the master and the other slave. The slave and master processes for a given transaction are determined at design time. In Figure 4.11, we assume that *P1* is the slave process and *P2* as the master process. Hence, *P1* sets the synchronization flag and *P2* tests the flag when it is ready to start the transaction. If the flag is already set, the data transfer is initiated. However, if the flag is not set, *P2* must wait for an event notification from *P1* to know when the flag is set. The event notification is needed if *P2* becomes ready before *P1*. *P1* notifies this event when it sets the flag. Once *P2* reads the flag as set, it recognizes that *P1* is ready and resets the flag. This completes the synchronization phase by which two processes prepare to transact data.

ARBITRATION MODELING

After synchronization, the master component will attempt to reserve the bus for data transfer. This is necessary since the bus may be a shared resource, and multiple transactions attempted at the same time must be ordered sequentially. The master process makes an arbitration request to the bus arbiter. The arbiter resolves multiple bus requests and allows only one master to access the bus at any given time. The simplest way of modeling the arbiter is to use a mutual exclusion object, commonly known as a mutex. An arbitration request corresponds to a mutex lock operation. Once the transaction is complete, the process releases the bus with a mutex unlock operation.

Although mutexes are convenient ways of modeling arbiters, the functionality of the mutex may differ significantly from a real bus arbiter. If multiple lock requests are made to a mutex, the resolution policy may depend upon the implementation of the mutex object in the respective programming language.

However, bus arbiters have different types of resolution policies depending on the given bus protocol. Therefore, the arbiter component must be modeled uniquely for each bus. Some of the common arbitration policies include fixed priority, round robin (RR), and first come first serve (FCFS). We can model each of these policies using priority queues and status flags for the master components on the bus. The timing for making the arbitration request and releasing the bus is fixed and may be obtained from the database. The run-time delays for request resolution is also computed by the arbiter model.

DATA TRANSFER MODELING

After synchronization and arbitration, the sender process is ready to write the data on the bus. Due to the limited number of bus signals, the data transfer takes place in several parts, known as bus cycles. In each bus cycle, a fixed number of bytes are transferred on the bus; the number of bytes is known as the bus word size. However, for functional modeling purposes in the TLM, we may assume that the entire data structure is transferred in a single transaction. This transfer may be modeled as a memory copy. Although, this does not result in an exact representation of cycle accurate bus protocol, it provides a close approximation. We may use analytical methods to compute the actual transfer time depending on the size of the data.

For each channel transaction, we know the size of the data to be written or read. The simplest approximation of transfer delay can be determined as the product of the bus cycle time and the number of bus words for the given data size. This delay holds true if the bus permits only single word transactions to occur at any one time. However, some buses may support more optimized modes, such as burst or pipelining. In order to understand these bus modes, we must consider the various phases of a bus cycle. A bus cycle, at its most basic, may be divided into an address phase and a data phase. For a burst transaction, the bus is reserved by the master for several cycles. Then, the addressing is done only once and the data words are transferred consecutively without repeating the addressing. In the case of pipelined transfer, the data phase of the first transaction may overlap with the address phase of the next transaction. These optimized transfer modes may reduce the total transfer time by almost half.

Using the methods described above, we can generate timed TLMs that provide estimation of both computation timing as well as communication timing for heterogeneous embedded designs [2].

4.3.5   TLM SEMANTICS

It is crucial to establish TLM semantics in any design flow for several reasons. If the TLM's objects and composition rules are defined clearly, it is possible to develop methods to automatically generate TLMs from the given mapping of an application to a platform. The TLM acts as a high speed executable model that encompasses all the system level design decisions. But the importance of TLM semantics is not limited to simulation; they also make it possible to synthesize cycle accurate models as well. These cycle accurate models include embedded software and RTL hardware for implementing the embedded system with standard CAD tools available in the market. Finally, well defined verification semantics allow the designer to check the functional equivalence of the TLMs with the original application model, making the designs generated from TLMs more reliable.

Figure 4.13 shows the TLM that has been generated from the mapping defined in fig:sys:SampleMap. Notice that both application and platform objects are captured in the TLM. We will describe the semantics of TLMs using SystemC, though they may be described in any system level design language. The PEs are modeled as modules at the top level in the SystemC representation. The buses are modeled as SystemC channels with well defined methods for common bus operations such as synchronization, arbitration, and reading/writing memory. Interfaces are also represented as SystemC modules with internal buffers modeled as FIFO channels. Memories are modeled as SystemC modules with arrays that are indexed by memory address [123].

To capture the software platform, the operating system (OS) is modeled as a channel instantiated inside the CPU. The OS channel provides common services like scheduling and inter-process communication (IPC), as well as critical section and timing functions. The hardware abstraction layer (HAL) is also modeled as a channel to provide an implementation of all the application level

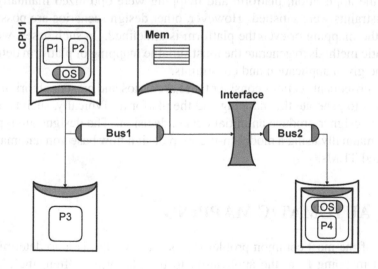

FIGURE 4.13 Automatically Generated TLM from system specification.

channels. The application processes are modeled as SystemC threads inside the CPU modules. The application threads, local to a CPU, may communicate using the communication services of the OS. Inter-PE communication takes place through the hardware abstraction layer, which in turn uses the respective bus channel methods. It must be noted that at the TLM abstraction level, we do not distinguish between hardware and software. This simplifies system level design because we can then use a uniform modeling style for TLMs.

Well defined semantics, such as those described above, allow for the automatic generation of a TLM from the given mapping of an application to a platform. There are several advantages to this. Clearly, the primary benefit of automatic TLM generation is that it saves time for model development and manual optimizations. The manual optimizations can also be made at a much higher abstraction, preferably using a graphical interface, and the TLM code can be generated automatically. Another advantage is that manually written models are not easily verifiable because it is difficult to establish correlation between objects in any two independently written models. Even if modeling rules are imposed, there are higher chances for human errors in following the rules for writing verifiable models. Automatically generated models, however, follow well defined rules, making it possible to correlate objects between any two models, This enables equivalence verification of application models and TLMs.

In this section, we considered a scenario in which the application model, platform definition and mapping were all specified. The TLMs were generated from the specification in order to evaluate the design choices. If the specified constraints were met, the decisions were finalized. If the constraint check failed, the application, platform and mapping were optimized manually until the constraints were satisfied. However, other design scenarios are possible in which the mapping or even the platform is undefined. In such cases, we need automatic methods to generate the most suitable mapping or platform definition from the given application and constraints.

The subsequent sections consider these scenarios and present algorithms and heuristics to generate the mapping and the platform. Typically, such heuristics help the designer produce an initial design decision. The designs are typically refined manually using a model-in-the-loop design flow based on automatically generated TLMs.

4.4   AUTOMATIC MAPPING

One of the most common problems in system synthesis is to determine the optimal mapping from the application to the platform. Often, the platform components are fixed, due either to legacy reasons or to availability. Only a few parameters of the platform may be configurable. Changing the platform entails huge overhead because it affects all the models as well as tool chain. Therefore, designers try to exercise their optimization options by modifying the mapping instead.

As we saw in the previous section, there are well defined rules for mapping application objects to platform objects. Even with those rules, however, there will be several mapping possibilities if there are several objects in the application and the platform. For instance, if there are N processes in the application, and M fully connected processors in the platform, the total number of possible mappings is $N!/(M! * (N - M)!)$ For only 6 processes and 3 processors, that turns out to be 20 possible mappings! For any large application and reasonably complex platform, it is infeasible to create and evaluate all possible mappings. Therefore, automatic mapping generation algorithms are needed at the system level.

This problem of determining the optimal mapping has long been studied in various contexts beyond system level synthesis. Often, the problem is formulated as a cost minimization consideration. In such approaches, a cost function is created to compute the *goodness* of a mapping solution. The problem is to find the minimal cost mapping solution from all possible mappings. Various evolutionary algorithms have been proposed to solve this problem, with mixed results. In the context of system synthesis, the fundamental bottlenecks are the high number of possible solutions and the extensive time required to evaluate the cost function or performance of the design. Therefore, the most practical ap-

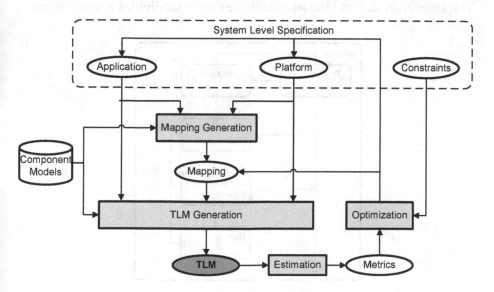

FIGURE 4.14 System synthesis with fixed platform.

proach to solving the mapping problem is to develop heuristics based mapping algorithms.

In this section, we will discuss some useful mapping heuristics and algorithms that operate on graph based representations of applications and platforms. First we will present an example application to demonstrate our mapping algorithms. We will describe basic techniques for application profiling and the creation of a weighted application graph. Then we will discuss two algorithms that take the application graph and platform as an input and produce a feasible application to platform mapping.

4.4.1   GSM ENCODER APPLICATION

In order to demonstrate our mapping and platform generation algorithms, we will be using a GSM encoder application [76]. The GSM encoder is an audio conversion application used widely in cellular phone designs, so it is an ideal representative design driver from the multimedia domain.

The GSM encoder application specification is illustrated in Figure 4.15 using a PSM model of computation. The top level encoder process is a hierarchical composition of five leaf level processes: LP_Analysis, Open Loop, Closed Loop, Codebook Search, and Update. The model also shows the data communication between the processes with thick straight edges. The thin curved edges show the control flow such as conditional execution and loops.

The input to the encoder is the raw speech data in commonly used wav format. This input is divided into frames, which are further subdivided into sub-frames.

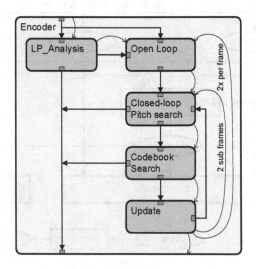

FIGURE 4.15   Application example: GSM Encoder

In the first step, LP Analysis, the parameters of a low pass (LP) filter are extracted. The contribution of the LP filter is then subtracted from the input speech. Next, using the past history of excitations, all the possible delay values of the pitch filter are searched for a match closest to the required excitation. The search is divided into an open-loop and a closed-loop search. A simple open-loop calculation of delay estimates is done twice per frame.

In each sub-frame, a closed-loop analysis-by-synthesis search is performed around the previously obtained estimates. This analysis is done in order to obtain the exact filter delay and gain values. The long-term filter contribution is subtracted from the excitation. The remaining residual comprises the input to the codebook search. For each subframe an extensive search through a fixed codebook is performed for the closest matching code vector. For each subframe, the coder then produces a block of 13 parameters for transmission. Finally, using the calculated parameters the reconstructed speech is synthesized in order to update the memories of the speech synthesis filters, reproducing the conditions that will be in effect at the decoding side.

4.4.2 APPLICATION PROFILING

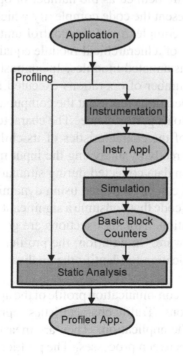

FIGURE 4.16 Application profiling steps.

Profiling is the first step in mapping generation. It analyzes the input model in terms of different metrics that characterize the application; these metrics will be used later for automatic decision making. To illustrate profiling, we will be using the execution delay of the application as the metric to be profiled. Application profiling is similar to timing analysis in TLMs, as discussed in Section 4.3.4. However, since the mapping is not available, the profiling measures only raw metrics such as the number of operations and data transactions.

A typical instrumentation-based profiling flow is illustrated in Figure 4.16. In an instrumentation-based profiling approach the application is first instrumented with counters that are incremented at the end of basic blocks. Then, a sample input is used to execute this instrumented model. At the end of the simulation, the counters provide us with an execution count for each block level in the application model. In the second step, the application characteristics are calculated by statically analyzing the code together with the collected execution counters per basic block. Specification characteristics are computed hierarchically for each process, channel and variable in the application model. The profiling characteristics are classified into three categories: Computation, Communication, and Storage [32].

In each category, we compute both static and dynamic metrics. Static computation characteristics are defined as the number of operations in the code of each module. They represent the code complexity which is related to code size and the implementation complexity of the control unit in general. The static operation characteristics of a hierarchical module equal the sum of the characteristics of all its child/instantiated modules. In contrast, dynamic computation characteristics are the number of operations executed by each module during simulation. Dynamic operations represent the computational complexity in the system which is related to its performance. The characteristics of a hierarchical module equal the sum of the characteristics of its child module. Static characteristics are derived directly by analyzing the input model whereas dynamic characteristics depend on data collected during simulation.

Application hot spots can be identified using dynamic computation metrics. Hot spots are sections of code that consume a significant amount of computation from the overall application. Hot spot sections are generally good candidates for acceleration in hardware. In addition, the profiler extracts the control information from the application to identify the applications call graph and code structure.

Besides computation, communication profile of the application is also needed to make mapping decisions. Traffic characteristics represent the complexity of the communication in the application. They are measured as the amount and type of data exchanged between processes. The profiler provides separate input and output traffic characteristics. As processes communicate through variables and channels connected to their ports, traffic characteristics are attached to process ports, and therefore also to the variables and channels that are connected to the ports. Furthermore, traffic characteristics are also attached to processes.

Static traffic characteristics are the number of connected ports of a certain type. They represent connectivity complexity, which relates to the message passing traffic incurred between two dependent processes in order to make the output of a process available at the inputs of the next process. In contrast, dynamic traffic characteristics are defined as the number of times a port or a variable/channel of a certain type is accessed during simulation. An access is generated whenever a statement in the code reads from a port variable, writes to a port variable, or calls a port interface method. Dynamic traffic characteristics represent access complexity which relate to the traffic incurred for a shared memory implementation of communication between dependent processes.

Figure 4.17 shows a selection of profiling results for our sample GSM encoder. We report metrics for computation in each process and the communication on each data dependence edge. For computation, the number of total operations is shown for each process in the encoder. The figure shows that the Codebook search, with 646 Million Operations, is the most compute intensive process.

The pie chart in Figure 4.17 shows the detailed profiling results for the Codebook process itself by breaking down the percentage of the total computations for each type of operation. This profile data helps us evaluate the demand in the number and types of functional units that can most optimally implement the process. We can see that for the Codebook search, most operations are integer multiplications followed by integer additions and a small percentage of integer divisions. Therefore, mapping the Codebook search process to a processor that has ALUs, multipliers and dividers to support the operations would be ideal.

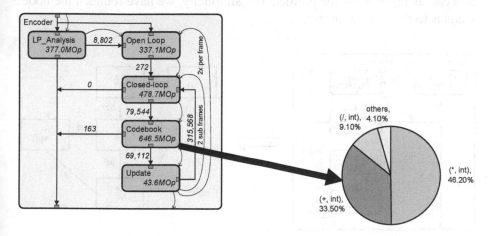

FIGURE 4.17 Profiled statistics of GSM encoder.

The labels on the communication edges quantify the amount of data that is transferred between the encoder processes. The profiling reveals high traffic between the modules closed loop and update (315kBytes). The Codebook has significant communication (79K with the closed loop, and 69K with the update module). This indicates that it would be preferable to map Closed Loop and Update to processors that are connected in the platform with a fast bus or link.

Overall, the computation and communication statistics collected during profiling provide us with a quantitative analysis of the computation and communication demands of the various processes and channels. Although, this data in itself is not sufficient to make optimal design decisions, it does help with creating a preliminary mapping, which may be refined later. We will now look at a few algorithms that can be used to automatically create a mapping from the application to the platform, based on the profiled statistics.

4.4.3   LOAD BALANCING ALGORITHM

The first mapping algorithm we will discuss is the load balancing algorithm. The basic idea of this algorithm is to go over a sorted list of processes and map it to the least loaded PE in the platform. But before we present the algorithm, we must formulate the problem using graphical representations of the application and the platform. Graph based representations will help us define the problem mathematically.

The application graph can be easily obtained from the profiled application as shown in Figure 4.18. The leaf level processes are translated into nodes. The node labels correspond to the process names as shown. We also assign weights to the nodes based on the number of operations (in millions) executed in the process, as reported by the profiler. For simplicity, we have rounded the node weights to the nearest integer.

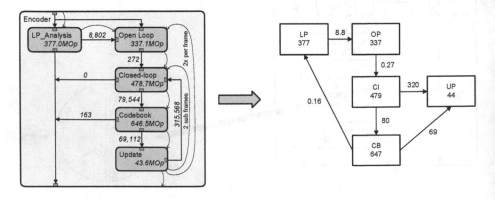

FIGURE 4.18   Abstraction of profiled statistics into an application graph.

Two nodes, say x and y, in the application graph are connected by a directed edge, in this case from x to y, if process x sends data to process y. The weight of the edge is the number kilobytes of data transacted, as reported by the profiler. As in the case of node weights, we have rounded the edge weights to the nearest integer for simplicity.

It must be noted that we have not taken control dependencies into account while creating the application graph because adding the control dependencies may make the application graph unduly complex. Furthermore, it would make the mapping algorithm significantly more difficult. Therefore, for simplicity, we are not considering the effects of concurrency when generating the mapping.

As with the application graph, we need to create the platform graph to simplify the platform input for mapping generation. The conversion from the platform definition to the platform graph is also very straightforward as shown in Figure 4.19. Each PE in the platform is represented as a unique node in the platform graph. Each node is assigned a weight, as they were in the application graph. The node weight is the computation speed of the particular PE type, measured in millions of operations per second.

The platform shown in Figure 4.19 has three PEs of three different types: *CPU*, *HW*, and *DSP*. The three PEs have different computation speeds, as shown by the node weights in the platform graph. The HW is assumed to be a specialized implementation of which ever process is mapped to it. Although the numbers are estimated, they are based on the general relative performance of the three types of processing elements.

The edges in the platform graph represent the possibility of sending data from one processing element to another as determined by the analysis of the platform for feasible communication paths. Recall that for direct communication on any segment, we need one component to be the master and the other to be the

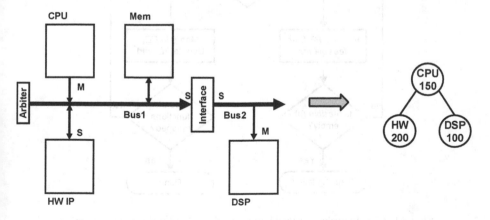

FIGURE 4.19 Creation of platform graph.

slave. We can see on the platform definition that *CPU* is the master on *Bus1* while *HW* is the slave. Therefore, a direct communication path exists between *CPU* and *HW*. The interface can be used for communication between *CPU* and *DSP*, via the *Interface*. This is true because the *Bus1 master CPU* can communicate to the *Bus1 slave Interface*. Similarly, the *Interface*, which is also the *slave* on *Bus2*, can communicate with the *Bus2 master, DSP*. However, no communication is possible between *HW* and *DSP*. *HW* cannot communicate with the *Interface* because they are both slaves on *Bus1*. Therefore, we do not add an edge between *HW* and *DSP* in the platform graph. Note that edges in the platform graph do not have any weights associated with them because we do not consider communication costs in the load balancing algorithm.

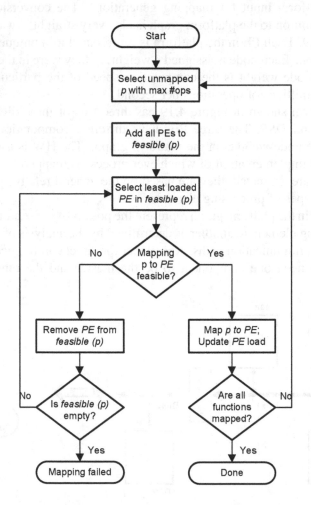

FIGURE 4.20 Flowchart of load balancing algorithm for mapping generation.

The purpose of using a load balancing algorithm is to create a many-to-one mapping from the nodes of the application graph to the nodes of the platform graph. The heuristic used to create the mapping attempts to evenly distribute the computation load for each PE in the platform. We define the computation load as the total number of operations divided by the speed of the PE.

Therefore, we can say

$$Load(PE) = \Sigma MOps(p)/Speed(PE), \; such \; that \; p \; is \; mapped \; to \; PE$$

Next we define the set of feasible mappings for a given process p. We say that processor *PE* is in `feasible(p)` if there does not exist any other process q, such that p is connected to q in the application graph and q is mapped to a processor that is not connected to *PE* in the platform graph.

Using the functions *load* and *feasible*, we may define the load balancing algorithm as shown in Figure 4.20. We start by selecting the most computationally intensive unmapped process, say p, to the least loaded PE. We initialize the feasible list by including all the PEs in the list. If there are unloaded PEs (without any processes mapped to it), then we select the fastest PE for mapping. Otherwise, we select the PE with the minimum load. Before mapping, we check if the mapping from p to *PE* is feasible. If it is feasible, we store the mapping, update the PE load and select the next unmapped process. However, if the mapping is infeasible, we remove the PE from the list `feasible(p)` and repeat the mapping attempt by selecting the next least loaded PE in the feasible list.

Let us illustrate the algorithm using the example application graph in Figure 4.18 and the platform graph in Figure 4.19, as inputs. We start by taking the most computationally intensive process, *CB*, and map it to *HW*, which is the fastest PE. Therefore, we have

$$Load(HW) = Mops(CB)/Speed(HW) = 647/200 = 3.24$$

Next, we pick process *CL* and map it to *CPU*. This gives us

$$Load(CPU) = Mops(CL)/Speed(CPU) = 479/150 = 3.19$$

CPU is, therefore, less loaded than *HW*, but *DSP* is still unloaded. Next, we consider process *LP* to map to *DSP*. However, *LP* communicates with *CB*, which is mapped to *HW*. Since there is no communication link between *HW* and *DSP*, we cannot map *LP* to *DSP*, so we must map *LP* to the next least loaded PE, which is *CPU*. As a result, we get

$$
\begin{aligned}
Load(CPU) &= [Mops(CL) + Mops(LP)]/Speed(CPU) \\
&= [647 + 377]/150 = 6.83
\end{aligned}
$$

Next, we consider the unmapped process *OP*. Since *OP* does not communicate with *CB*, it is feasible to map it to *DSP*, giving us

$$Load(DSP) = Mops(OP)/Speed(DSP) = 337/100 = 3.37$$

Finally, we can map the last process *UP* to the least loaded PE, which is *HW* because *UP* does not communicate with *OP*. As a result, the algorithm terminates with *CL* and *LP* mapped to *CPU*; *CB*, *UP* mapped to *HW*; and *OP* mapped to *DSP*.

Although the load balancing algorithm uses a reasonable heuristic, there are drawbacks to using it. For example, it is possible that the algorithm may not terminate with a mapping solution. It is very sensitive to the order in which processes are selected for mapping. Mapping a set of processes may make it impossible to map the remaining processes. More crucially, this algorithm does not take communication into account while creating the mapping which may lead to poor design for communication intensive applications. We will next consider a mapping algorithm that attempts to minimize the overall execution time while taking communication delays into account.

4.4.4   LONGEST PROCESSING TIME ALGORITHM

Depending on the application and the mapping decisions, the communication delay may become significant. In such cases, we cannot ignore the effects of communication mapping on the overall execution time. The key drawback of the load balancing algorithm is that it does not take communication timing into account; it only considers the feasibility of communication. Therefore, designers need a new heuristic for mapping decisions that accounts for communication cost along with computation cost.

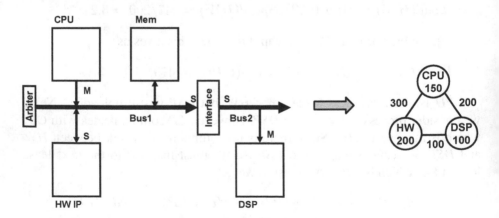

FIGURE 4.21   Platform graph with communication costs.

The algorithm we will present here to meet this need is called the longest processing time (LPT) algorithm. In multiprocessor scheduling, several variants of this algorithm have been used to map tasks in a program to a set of processors [108]. The same principle can be used to solve the mapping problem at the system level. The key idea in LPT is to choose the processes for mapping in decreasing order of their number of operations (hence the name). Furthermore, this algorithm computes the cost function for mapping a process *p* to each *PE*, called C (p, PE). The process *p* is then mapped to the *PE* at the minimal cost. The computation overhead of mapping is also taken into account by the LPT algorithm. Hence, we need additional communication costs in the platform graph that will help determine the best mapping of inter-process communication to buses and routes in the platform.

Figure 4.21 shows a platform graph with communication costs. The platform, which we will use as an illustrative example, is derived from the platform in Figure 4.19. In the original platform, we did not have a communication path from *HW* to *DSP*, because *HW* was a slave on *Bus1*. In order to create a communication path from *HW* to *DSP*, we have added a direct memory address (*DMA*) component on *Bus1*. The *DMA* is a master on *Bus1*, as shown, so it can communicate with both *HW* and *DSP*, via the *Interface* component. The *HW* can send the data to the *DMA*, which in turn can forward it to the *DSP*.

The edge weights reflect the connection speed between the processors. Given nodes *x* and *y* in the platform graph, the weight of edge *(x, y)* is the effective speed of data transaction between PEs *x* and *y* in kilobytes per second. Since *CPU* and *HW* are connected directly on *Bus1*, they have the fastest transaction speed. The data from *CPU* to *DSP* must be buffered at the *Interface*, making the transactions between the two slower. Finally, transactions between *HW* and *DSP* must go though *DMA* and *Interface*, so they are the slowest in the platform.

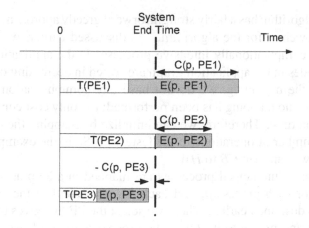

FIGURE 4.22 LPT cost function computation.

Before we delve into the LPT algorithm, we must first introduce the mapping cost function that the algorithm uses. Figure 4.22 shows a snapshot of a step in the mapping algorithm. It assumes that we have three processing elements in the platform, *PE1*, *PE2* and *PE3*. In this step of the algorithm, we are attempting to map process *p* to one of the PEs. The X-axis shows the linear time. Variable *T(PE)* represents the current execution end time for the PE. In other words, this is the time taken by the PE to execute all the processes mapped to it. The system end time is the maximum *T(PE)* for all the PEs in the platform, since all the PEs are assumed to be executing in parallel.

Variable *E(p, PE)* keeps the time estimated to execute process *p* on *PE*. This includes the computation time as well as the total communication time. The computation time is calculated simply by dividing the weight of node p in the application graph by the weight of node *PE* in the platform graph. The communication time is assumed to be negligible for all the local inter-process communication inside *PE*. If the communicating processes are mapped to different PEs, we calculate the communication time by dividing the size of communicated data by the communication speed. More precisely, assume we are given processes *p1* and *p2*, mapped to *PE1* and *PE2*. The communication time between *p1* and *p2* would then be the weight of edge *(p1, p2)* in the application graph divided by the weight of edge *(PE1, PE2)* in the platform graph. The total communication time for process *p* is calculated by adding all the communication times with neighbors of *p* in the application graph.

The cost of mapping *p* to *PE*, C(p, PE) is the extra time added to the system end time as a result of the mapping. This time is illustrated for the three possible mappings in Figure 4.22. The time may be negative as in the case of C(p, PE3). So we can say that

$$C(p, PE) = T(PE) + E(p, PE) - SystemEndTime$$

The LPT algorithm has a fairly straightforward greedy approach. Figure 4.23 shows the flowchart for the algorithm. As discussed earlier, we start with a sorted list of computationally intensive processes in the application. In other words, the nodes in the application graph are sorted in decreasing order of their weights. At the beginning, we do not have any communication costs to be measured since no mapping has been performed; the only cost considered then is computation time. Therefore, we can initialize by mapping the process with the highest number of operations to the fastest PE. So in the example shown in Figure 4.18, we can map *CB* to *HW*.

The remaining unmapped processes are mapped in a loop as shown in the flowchart. For each process *p*, and each *PE*, we calculate the cost function C(p, PE) as described earlier. Then, we select the PE that gives the minimum cost function and map *p* to it. Once the process is mapped, we must update the execution end time for all the PEs as well as the overall system end time.

This is because the variable *T(PE)* may be modified if a process mapped to PE communicates with *p*. Once all the *T(PE)* values have been updated, we can obtain the system end time. The cost function computation, mapping and timing updates are repeated until all the processes have been mapped. The solution for mapping the application in Figure 4.23 to the platform in Figure 4.21 can be produced simply by following the algorithm steps; we have left this as an exercise for the reader.

Since the LPT algorithm uses a simple greedy heuristic, it is computationally very efficient. Given an application with N processes and a platform with M PEs, we can sort the processes in $O(NlogN)$ time. The cost function for each PE can be calculated in $O(N)$ time, because we must inspect if the communicating processes for a chosen process p are mapped to another PE. Therefore the cost computation for all PEs can be done in $O(N*M)$. Finally, the

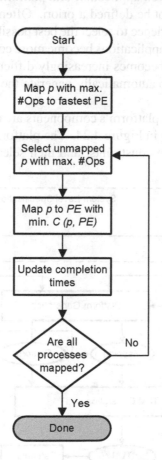

FIGURE 4.23 Flowchart of LPT algorithm for mapping generation.

loop is executed N times, which gives us a total time complexity of $O(N^2 * M)$. Therefore, the LPT algorithm runs in polynomial time.

In this section, we discussed algorithms for mapping a given application to the platform. However, in some system design cases, the platform itself may not be given. In the next section, we will discuss such a design scenario and present heuristics and algorithms for constructing a platform from the given application graph.

4.5   PLATFORM SYNTHESIS

In the previous section, we discussed the automatic generation of mapping decisions based on the given application and platform. In some design cases, the platform itself may not be defined a priori. Often, designers use common knowledge and their experience to select the best possible platform for the given application. However, as applications become more complex, designing the optimal platforms for them becomes increasingly difficult. Designers, therefore, need methods and tools to automatically generate the optimal platform from a given application.

The characteristics of a platform's components are defined in the component model database, as shown in Figure 4.24. The platform generator analyzes the application and the design constraints, and then selects components from the

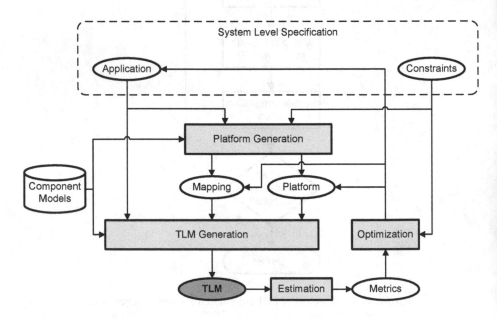

FIGURE 4.24 System synthesis from application and constraints.

database to be instantiated in the platform. Since the platform components are chosen based on the application, the mapping is implicit. For example, if a process *p* in the application is best executed by component PE, then an instance of PE is added to the platform. Therefore, it is implicit in the selection of PE that process *p* will be mapped to it.

The platform selected by the platform generator may not be the most optimal; however, in the absence of any given decisions, the generated platform may be suitable to at least initialize the process of design optimization. For optimization, we can use the same model-in-the-loop approach that we have used to this point. The application and the automatically generated platform and mapping are used to generate the TLM of the system. The TLM may be simulated to obtain metric estimates. These estimates may be compared to the constraints, which lead to an optimization of the application, platform, or mapping. The optimization loop is shown with thicker arrows in Figure 4.24.

TABLE 4.1 A sample capacity table of platform components.

PE Type (Cost)	Speed	Capacity (*3 sec)
CPU (2)	100	600
DSP (1)	50	300
HW (5)	200	1200

4.5.1   COMPONENT DATA MODELS

As we mentioned earlier, we use the component data models, along with the application and constraints, to generate the platform. A simple example of one part of a component data model is shown in Table 4.1. We consider three types of components here, namely *CPU*, *DSP*, and *HW*. The two given characteristics of each component are the relative cost and speed. The cost may refer to the dollar amount for purchasing the IP or the development cost for building it. *DSP* is the cheapest component followed by a general purpose embedded *CPU*. *HW* is the most expensive component because of the significant effort needed to develop and verify it; though the speed of the component is typically inversely proportional to its cost. That is to say, custom hardware may be expensive to develop, but it executes the process much faster than a *CPU* or *DSP* would.

The last column in Table 4.1 specifies the capacity of the components. By capacity we refer to the millions of operations that can be performed by the component under the given timing constraints. For illustration purposes, the timing constraint has been defined as 6 seconds. Therefore the system to be designed must complete the execution in less than 6 seconds. The capacity figures for the PEs are obtained by multiplying the speed of the PEs with the timing constraint, which again is 6 seconds. Note that the capacity numbers cannot be stored in the database. They must be computed based on the given timing constraint. Similar capacity numbers can be used for other metrics such as energy and bandwidth.

4.5.2   PLATFORM GENERATION ALGORITHM

Before we discuss the details of the platform generation algorithm, we must introduce a few terms that are used in the algorithm. As in the previous algorithms, we consider the computational cost of the process to be the weight of the process node in the application graph. We define the available computation capacity in the PE as $Slack(PE)$. Therefore, we can say

$$Slack(PE) = Capacity(PE) - \Sigma MOps(p), \; for \; all \; p \; mapped \; to \; PE$$

The platform generation algorithm also considers the communication while generating the implicit mapping. If a given process p has a high amount of communication to the already mapped processes in *PE*, then the algorithm attempts to map p to *PE*. The reasoning is that communication costs the least if it is local to a processor. Therefore, to take communication into account, we define the closeness factor $C(p, PE)$ between process p and processor *PE*. In context of the application graph, we define $C(p, PE)$ as the sum of all the weights of the edges between p and its neighbors mapped to *PE*. So we can say

$$C(p, PE) = \Sigma Comm(p, q), \; for \; all \; q \; mapped \; to \; PE$$

Figure 4.25 shows the flowchart of a greedy algorithm for platform generation. The basic principle is to traverse the set of processes and map them to the *closest* processor or the one with the maximum slack available. If no such processor is found, the processor that can execute the given process at the lowest cost is selected from the database and instantiated.

We start with a list of application processes that is sorted in decreasing order of computation cost. Clearly, the first process, p, cannot be mapped anywhere since no processor exists in the platform yet. It is therefore mapped to the least cost processor, say PE, such that $Capacity(PE) \geq MOps(p)$. After the initialization, the remaining processes are mapped in the main loop. There are two checks performed, as mentioned earlier. The closeness factor is considered first for mapping, failing which, the available slack is considered. It must be noted that the computational cost of any process should not be higher than the capacity of all the processors in the database. In such a case, a platform cannot be found. If this is the case, the designer may want to modify the constraint, split the application, or add faster components to the database. Once all the processes are mapped, the platform is generated and the algorithm terminates. For the sake of simplicity, we will consider a single bus platform. If PEs with incompatible interfaces are instantiated, then multiple buses may be instantiated and connected with an interface.

We will illustrate the greedy platform generation algorithm using the GSM voice encoder application, whose application graph is repeated here in Figure 4.26(a). We will be using the component database shown in Table 4.1.

The first process selected is *CB* with 647 MOps. The only processing element with the capacity to execute *CB* under the given 6 second constraint is *HW*. Therefore, we add *HW0* as an instance of *HW* to the platform, and map *CB* to

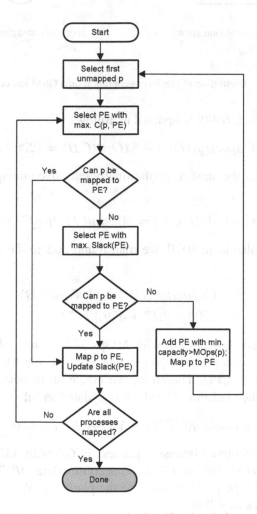

FIGURE 4.25 Flowchart of a greedy algorithm for platform generation.

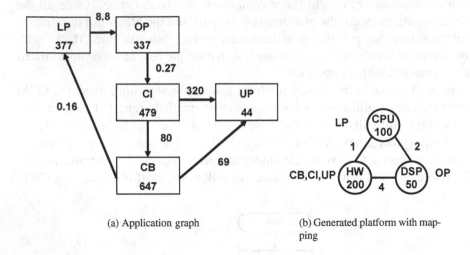

(a) Application graph

(b) Generated platform with mapping

FIGURE 4.26   Illustration of platform generation on a GSM Encoder example.

it. The variable *Slack(HW0)* is updated as follows

$$Slack(HW0) = Capacity(HW) - MOps(CB) = 1200 - 647 = 553$$

Next we consider the most computationally intensive unmapped process *CI*. We can see that

$$C(CI, HW0) = Comm(CI, CB) = 80, \ and \ MOps(CI) < Slack(HW0)$$

Since *CI* may also fit in *HW0*, we map it and update the slack of *HW0* as follows

$$\begin{aligned} Slack(HW0) \ &= \ Capacity(HW) - [MOps(CB) + MOps(CI)] \\ &= \ 1200 - [647 + 479)] = 74 \end{aligned}$$

Next we consider process *LP* with MOps 377. Since *LP* cannot fit into HW, we select the cheapest component that can execute LP under the given constraints, which is CPU. Therefore, we create an instance of CPU called CPU0 and add it the platform. The slack is updated as follows

$$Slack(CPU0) = Capacity(CPU) - MOps(LP) = 600 - 377 = 223$$

The next most compute intensive process is *OP* with 337 MOps. Again, we find that neither *CPU0* nor *HW0* can accommodate *OP*. Therefore, a new instance *CPU1* of type *CPU* is added to the platform. We map *OP* to *CPU1* and update its slack as follows

$$Slack(CPU1) = Capacity(CPU) - MOps(OP) = 600 - 337 = 263$$

Finally, we are left with process *UP* which has only 44 MOps, so we can map it to any of the processors in the platform. However, based on the closeness heuristic in the greedy algorithm, we must map *UP* to the PE with which it has the most communication. The closeness factors are computed as follows for each PE

$$
\begin{aligned}
C(UP, HW0) &= Comm(UP, CB) + Comm(UP, CI) \\
&= 69 + 320 = 389 \\
C(UP, CPU0) &= Comm(UP, LP) = 0 \\
C(UP, CPU1) &= Comm(UP, OP) = 0
\end{aligned}
$$

Therefore, we must of course map *UP* to *HW0*. As a result, we get the mapped platform graph shown in Figure 4.26(b). Consequently, there are three components in the platform connected with a common bus, since both *CPU* and *HW* type components can be connected to the same bus. This is an illustrative example of how platform generation can be done automatically using a greedy algorithm.

To sum up, in this section, we discussed the methods for automatic platform generation, as well as how to implicitly create the mapping. The system synthesis algorithms, discussed so far, allow the designer to make some useful system level design decisions. However, once the TLM is generated, a back end synthesis flow is needed to generate the final implementation model. The next section provides an overview of the back end flow, which is then discussed in greater depth in the following chapters.

4.5.3 CYCLE ACCURATE MODEL GENERATION

So far, we have discussed the system level design decisions required to create a platform and map the application to the platform. We saw how the TLM can be used to evaluate the system level design decisions. However, the system level decisions, in most cases, are still too abstract and incomplete to implement the design in that state. In order to implement the design, a synthesizable pin-cycle accurate model (PCAM) of the system must be created. This model is supported by traditional FPGA and ASIC design tools for manufacturing. The crucial point is that if we have created the TLM according to well defined synthesis semantics, we can easily generate the CAM from the TLM.

As shown in Table 4.1, there are three steps to CAM generation; software synthesis, hardware RTL synthesis and interface synthesis. For software synthesis, a RTOS and HAL library is used to automatically generate the system software stack for each software processor in the design. This system software is application and platform specific; it presents a programming interface to the application processes which is identical to the TLM's. The application code and the system software can be compiled into a single binary for download to the specific SW processor.

If there are custom hardware components in the design, it is possible to generate hardware C models that can be synthesized into equivalent RTL models using traditional high level synthesis tools. The RTL output of these tools is integrated back into the CAM. Alternately, a pre-designed RTL IP component may be used directly in the CAM. Finally, interface synthesis tools can be used to automatically synthesize RTL hardware description of all interfaces between buses in the platform. The interface synthesis tools need a cycle accurate model of the relevant bus protocols, which may be obtained from a library.

The generated CAM can be used for either an ASIC flow or FPGA flow. Commercial ASIC design tools for logic and physical synthesi can be used to produce the layout from the CAM. The CAM can also be used for logic simulation at the cycle level. Alternately, commercial FPGA design tools may be used for prototyping the design from the CAM. Therefore the CAM, which is synthesized from the TLM, can be used for multiple types of implementation.

4.5.4 SUMMARY

In this chapter we discussed the methodologies and techniques for system synthesis. As the abstraction level of design and modeling is raised beyond the traditional cycle accurate level, system synthesis will become a neccessity,

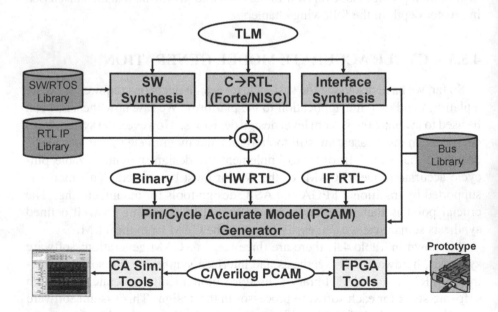

FIGURE 4.27 Cycle accurate model generation from TLM.

not a luxury. Virtual platform based development is a first step in realizing the goal of automatic system synthesis. While virtual platforms allow the concurrent development of hardware and software, we advocate that a model based methodology, which improves on the promise of virtual platforms, be developed with the help of system synthesis tools.

To this end, we discussed system synthesis in the context of TLMs, showing how TLMs can be automatically generated from a given mapping of the application to a platform. We delved into the details of timed TLM generation for evaluating the performance of a system. The performance data obtained from the TLM execution can be used to check for constraints and to optimize the design to meet those constraints. We also looked at methods to automatically create a mapping if no initial mapping is available, and discussed algorithms and heuristics for automatic synthesis of the platform from a given application model and library of components.

Although this approach permits us to make several critical design decisions at the system level, we need CAMs for the final system implementation and manufacturing. If the TLM semantics are well defined, they can serve as a starting point for CAM generation. In the following chapters, we will discuss the principles of embedded software synthesis and the hardware RTL synthesis methods which enable the generation of this CAM.

Chapter 5
SOFTWARE SYNTHESIS

This chapter describes software synthesis. As discussed in the previous chapter and shown in Figure 5.1, software synthesis, together with hardware synthesis and interface synthesis, is part of the component synthesis. We should recall that system synthesis produces a system model to describe the system's components and their communication. The bottom portion of the flow uses the system model as an input and generates an implementation for each component. Hardware synthesis, which we describe in the next chapter, generates an RTL description of custom hardware components. Software synthesis, topic of this chapter, produces binary code for programmable processing elements.

Software development dominates the design cost of modern complex multi-processor systems. The amount of software embedded in designs is increasing,

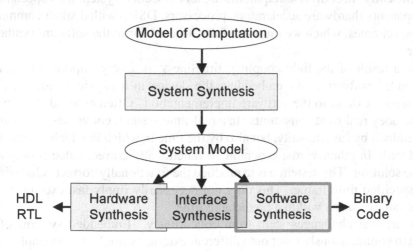

FIGURE 5.1 Synthesis overview

D.D. Gajski et al., *Embedded System Design: Modeling, Synthesis and Verification*,
DOI: 10.1007/978-1-4419-0504-8_5,
© Springer Science + Business Media, LLC 2009

partly due to increased design complexity, but also as a result of a shift toward software-centric implementations, which designers prefer because software allows them to flexibly and efficiently implement complex features. While embedded software was traditionally implemented manually, this method is too time consuming to meet today's time-to-market requirements. The extensiveness of implementation detail causes long development times, especially as embedded software is tightly coupled to the underlying hardware. Manual code development is tedious and error prone. To increase productivity, a method for automatically generating embedded software is much preferred.

5.1   PRELIMINARIES

The main challenges for developing embedded software stem from its tight coupling to the underlying hardware and external processes. This coupling may be at different levels. At one level embedded software drives and control customized hardware accelerators that are integrated to the platform. At another level, the complete embedded system is often part of a physical control process, such as an anti-lock brake system. There, embedded software can implement a control loop of the physical process though reading from sensors (e.g. measuring tire rotation speed) and controlling of actuators (e.g. setting the applied brake pressure). In both cases the software is specific to the underlying hardware. Specialized software drivers are necessary to access the custom hardware components (such as accelerators, sensors, and actuators). Already this small view into an embedded system indicates challenges of embedded software that is tightly coupled to a heterogeneous underlying hardware platform. In order to efficiently meet diverse requirements, an embedded system uses specialized components (hardware accelerators, processors, DSPs) with distinct communication schemes, which we must carefully account for in the software synthesis stage.

As a result of the tight coupling, timeliness is a very important aspect in embedded software. By embedding the device in a physical process, time constraints extend to the software implementation. Often embedded software has to obey real-time constraints. In a real-time system, correctness is not only determined by functionality, but also by the time in which in which the result is produced. In other words, generating a functionally correct value is only half of the solution. The system has to produce the functionally correct value within the specified time frame. This does not necessarily imply fast execution, but rather an execution within a predictable time.

Additional challenges stem from concurrency. Embedded systems often have to simultaneously react onto different external stimuli. For example, consider a telephone line card that provides your home with telephone service and

ADSL for fast Internet access. Its processor runs many tasks in parallel: it monitors the line for any phone activity (e.g. off-hook), manages the ADSL connection, keeps track of performance-monitoring statistics, communicates with the central controller, in addition to managing many other tasks. Concurrency makes software development challenging in expressing concurrent algorithms and maintaining safe communication between flows of execution. More so, the concurrent execution of software poses challenges as it requires alternating between flows of execution (e.g. by using an operating system), which complicates maintaining real-time constraints.

Naturally, embedded software has to obey resource constraints in order to meet the product requirements. Constraints limit, for example, memory consumption, available computing power, consumable energy, power dissipation, and other resources.

A systematic approach to embedded software development is needed to address the overlapping challenges of coupling, timeliness, concurrency, and resource constraints. Software synthesis is one possible solution. Before introducing the steps involved in software synthesis, we will touch on some requisites, including programming languages for embedded systems, as well as overviews of real-time operating systems and traditional embedded software development.

5.1.1   TARGET LANGUAGES FOR EMBEDDED SYSTEMS

Embedded systems can be programmed in a wide range of languages such as assembly, C, C++, and Java. They differ in many aspects, for example in the abstraction level at which the program is written and with that the granularity of control over the processor.

Assembly (e.g. [192]) is the lowest level language from the set above. This language is basically a symbolic representation of the processor's machine code and only minimally abstracts the processors complexity. It provides very fine-grained control of the processor internals. For example, assembly allows direct control of the processor registers and instructions. However, with this fine-grained control, assembly programs are very verbose, which makes it unsuitable for larger projects. As a further complication, the assembly language is specific to the processor's Instruction Set Architecture (ISA), so two processors may use a different assembly. The differences in assembly languages significantly increase the learning effort when developing code for different processors.

The C language [106] provides a higher abstraction than assembly. It is a general purpose programming language, designed for easy compilation to efficiently map language constructs to the processor's assembly. It provides low-level features such as bit-wise activities and also allows direct access to memory management. Furthermore, C requires minimal run-time support, by which we mean the support code necessary for running any program. Although C offers low-level features, it is mostly independent from the processor's architecture. Portably written standard compliant C code can be compiled to a variety of processors. To ensure portability, the code has to follow coding guidelines. MISRA [8] is one example of a C language coding standard for portable applications in the automotive domain. Overall, the higher abstraction level of C make this language more convenient and efficient for development than assembly.

C++ [173] is a general purpose language which began as an enhancement to C. C++ is backwards compatible with C, retaining C's fairly low-level features while adding higher-level concepts, such as classes, inheritance virtual functions, and operator overloading. These higher level constructs facilitate object-orientated programming, which make C++ attractive for large projects. C++ is used in a range of applications, from embedded systems to large desktop applications. Typically only a subset of C++ is used in embedded systems to avoid large runtime overhead and to ensure efficient execution on the target. [44] shows an example of coding guidelines defining for embedded system programming.

Java [54] was created to meet a different set of criteria: portability, reliability, and longevity. Java can be described as a simplified derivative of C++. It omits many of the more complicated C++ constructs, including templates, namespaces, multiple inheritance, and operator overloading. Java can also hide complexities such as memory management and pointer arithmetic from the user. On one hand, these simplifications limit programming flexibility, but on the other hand, they prevent unnecessary mistakes caused by developers' misemployment of that flexibility.

Java was originally implemented with an interpreter to make its object code, the Java byte code, portable. This interpreter, the Java Virtual Machine or JVM, is part of the language, so a compiled Java program should run on any machine/-platform with a JVM, ensuring Java's portability. However, a purely interpreted approach like this has the drawback of slow execution speed. To overcome this speed limitation, Java programs can also be compiled for a specific processor. Such approaches may for example use ahead-of-time compilation, much like the traditional C++ compilation process, or just-in-time compilation, in which the Java code is compiled while executing on the target processor itself. To further improve speed, some embedded processors now support Java accelerators in addition to their native-machine code (for example ARM's Jazelle technology [159]).

The Java language contains constructs for concurrency and communication. Its standardized runtime environment, provides interfaces for thread management, communication and synchronization primitives. To hide the complexities of manual memory management, it utilizes an automatic garbage collector to recover unused memory blocks. However, the original Java specification does not facilitate real-time computing, with the main challenge being the garbage collection. To address real-time requirements, the Real-Time Java Specification [17] provides vital extensions to the Java language, for example defining predictable memory allocation schemes.

While each of these languages had unique capabilities and drawbacks for coding embedded systems, C is currently the one used most predominantly, though C++ and Java have gained popularity. With the quality of C compiler results and the increasing complexity of projects, assembly language is rarely used for an entire application. Instead, developers use assembly for special code that requires direct access to the processor registers (for example within processor startup, or for context switching), or for hand optimizing small timing critical code sections. The versatility and popularity of C as a target language for programming embedding systems recommends it as our focus for this chapter.

5.1.2   RTOS

The languages assembly, C and C++ do not inherently provide support for concurrency. Concurrency, however, is an integral part of embedded systems. To enable concurrent execution, embedded applications are often executed on top of an Real-Time Operating System (RTOS), which makes an RTOS an essential part of the embedded software.

Like a general purpose operating system, an RTOS is a software layer above the bare processor that controls the concurrent execution of applications and provides various services for communication and synchronization.. An RTOS differs from a general purpose operating system in many aspects. The most predominant difference is in timing behavior. A general purpose OS is typically geared toward fairness, to give all running processes a fair amount of execution time. With an RTOS, on the other hand, the goal is predictability, to enable timely execution within predicable bounds.

A real-time operating system is an operating system that facilitates the construction of real-time systems. We use the word "facilitate" because the RTOS by itself does not guarantee real-time behavior. It provides its services within a predicable time and offers predictable algorithms for scheduling flows of execution. Therefore, properly designed software together with a properly configured RTOS enables the construction of a real-time system. We distinguish between two categories of real-time systems, hard and soft real-time systems, based on the potential consequence of missing a deadline. In a hard real-time system, missing a deadline may lead to catastrophic consequences to the controlled environment. Examples of this type of real-time system include plant control and medical applications. In a soft real-time system, missing a real-time constraint is tolerable if overall service quality remains acceptable. Media applications typically fall into this category.

An RTOS offers a wide range of essential system services. An RTOS provides services for task management as well as for Inter Task Communication (IPC). Task management creates, terminates, and controls tasks, while IPC enables tasks to exchange information and synchronize with each other (e.g. mailbox, queue and event). An RTOS offers services to control resource sharing (e.g. mutex, and semaphore). Memory management is another important embedded system programming. To this end, the RTOS provides deterministic mechanisms for memory allocation. An RTOS also provides timing support for a timed execution (e.g. for periodic triggering of tasks or timeout support). Finally, an RTOS enables communication with external devices and offers interrupt management. In some cases, an RTOS can provide a diverse set of standard drivers, such as IP communication stack, flash management drivers, or file system support.

RTOS implementations come from a variety of sources. Proprietary commercial kernels include, for example, QNX Neutrino RTOS and WindRiver's VxWorks. There are also a number of free RTOS implementations such as eCos, RTEMS, and uC/OS-II. Others are soft real-time extensions of time-sharing operating systems, like for example RT-Linux and Windows CE. There are also numerous specialized and research RT kernels. Each of these RTOS implementations has own unique features, common for all of them is enabling concurrent execution within an embedded system.

An RTOS's ability to manage its various responsibilities is dependent on supported scheduling algorithms. In general, a scheduling algorithm determines the order in which different flows of execution are given access to a resource. Hence, the scheduling algorithm decides in which order to execute tasks that are ready for execution. Scheduling algorithms can be characterized according to various properties. [30] describes and analyzes scheduling algorithms in detail. Some classification properties for scheduling algorithms are:

Preemptive / Non-preemptive characterizes whether a task can be interrupted in the middle of its execution. When using a preemptive scheduling algorithm, a running task may be interrupted in its execution at any point in time according to the scheduling policy. With a non-preemptive algorithm, a task may not be interrupted within its execution and scheduling may only occur when operating system services are invoked. As a result, a task once started will execute until completion if it does not call any operating system services.

Static / Dynamic refers to whether task scheduling parameters can be updated during runtime. When using a static algorithm, the task scheduling parameters are fixed once the task is released. In a dynamic approach, such parameters may change during the lifetime of the task.

Off-line / On-line characterizes when scheduling decisions are made. In an off-line approach, the complete schedule for all tasks is determined before releasing any task, and is stored and executed by a task dispatcher at runtime. In an on-line approach, scheduling decisions are made at run-time, as each task is released.

Off-line scheduling algorithms are often used in hard real-time systems. Using a pre-defined schedule created by an off-line algorithm significantly eases real-time system analysis and allows deterministic execution since every execution sequence is known before hand. Furthermore, executing from a pre-defined schedule minimizes runtime overhead as no scheduling decisions are made at runtime. However, an off-line algorithm can be brittle and inflexible as it can only handle what is completely known before starting execution. On-line algorithms, on the other hand, allow flexibly adjusting to changes in the system, however at the cost of higher runtime overhead. On-line algorithms are widely used in current RTOS implementations. Examples of the scheduling policies typically available in an RTOS are:

Priority-based scheduling. In priority-based scheduling, the task order is determined by task importance. The designer assignees a priority to each task defining its importance. At any given time, the highest priority task from the set of ready to run tasks is selected. Priority-based scheduling is often used, as it is flexible and easy to implement. Other scheduling policies can be implemented by priority distribution. Priority-based scheduling is a online approach, typically used with a preemptive policy. Dynamic priority changes may be allowed during task execution.

Earliest Deadline First (EDF). In an EDF schedule, the task with the earliest deadline is scheduled first. The on-line algorithm requires task deadlines to be available during execution time. Each ready task, is added to a priority queue based on its deadline expiry. The process that is closest to its deadline is dispatched to the processor.earliest deadline first

Rate Monotonic (RM). In an RM schedule, tasks are assigned priorities in descending order according to the length of the period, in which the task with the shortest period will be assigned the highest priority. Therefore, a frequently running task gets preference over rarely running tasks.

Round Robin (RR). RR assigns tasks to the CPU based on time slices and tasks take turns. Using an RR schedule, each task is assigned a fixed time slice which defines the amount of processor time the task may take to execute contiguously. After a task has used up it's time slice it is put into the back of the scheduling queue and the RR scheduler assigns a new task to the processor. After all ready tasks had their chance to execute on the processor, the original task can execute again. RR scheduling emphasizes fairness, as each task is guaranteed a predefined amount of processor time. Some operating systems offering priority-based scheduling use an RR schedule for tasks with identical priority (e.g. WindRiver VxWorks).

The list above outlines a small selection of scheduling algorithms. A more comprehensive description can be found in [30].

5.2   SOFTWARE SYNTHESIS OVERVIEW

Software synthesis deals with programmable components, such as processors. It uses information captured in the system TLM to generate the embedded software and to produce a complete binary for each programmable component in the system. If all code leading to the final binary is generated automatically, we can eliminate the tedious and error-prone process of manual code writing. In addition, an automatic generation demands less processor-and-platform specific knowledge from the designer, and hence enables the designer to target a wider range of architectures. Moreover, software synthesis reduces the effort required for system validation because each synthesis step can be individually verified, thus reducing the validation effort for the system as a whole. Overall, automatic software synthesis significantly increases productivity by reducing the time necessary for the development and debugging of software code.

Figure 5.2 shows a more detailed flow for software synthesis. It uses the system TLM, which reflects system-wide architecture decisions, as an input and generates a target binary for each core. Software synthesis is divided into code generation and Hardware-dependent Software (HdS) generation. Code generation produces flat C code out of the hierarchical model captured in the SLDL. It converts module hierarchies into a set of C functions. Instance-specific variables are translated into a set of data-structure instances. Additionally, code generation also resolves the connectivity between modules into flat C code.

The second component of software synthesis, HdS generation, produces all the drivers and support code necessary to execute the above-generated C code on a given hardware platform, in particular, the critical aspects of multi-tasking, internal communication, external communication, and binary generation. HdS generation addresses the issue of multi-tasking to concurrently execute tasks on the same processor, typically by utilizing an off-the-shelf RTOS to schedule the tasks. HdS generation also uses this multi-tasking solution to manage the internal communication, which is the information exchange between tasks on

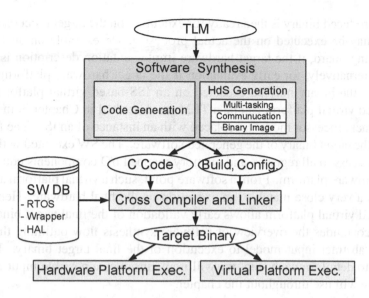

FIGURE 5.2 Software synthesis flow

the same processor. The most critical aspect of HdS, however, is external communication, for which it generates drivers so that a task on the processor can communicate with other processing elements. This includes synchronization with external components via polling, interrupts, or other methods.

After generating the task code and the supporting HdS, the final aspect is binary image generation, which procures the target binary to execute on the processor. Binary generation is a two stage process. First, HdS generation creates build and configuration files and second, a standard cross compiler and linker, directed by these build and configuration files, produces the final target binary.

HdS generation produces build and configuration files that control the build process (e.g. Makefile), which select and configure database components. As and example, the configuration files select an RTOS implementation from the database and configure it for execution on the selected processor. They also select specifically for the target platform a Hardware Abstraction Layer (HAL), which consists of low-level drivers for the timer, programmable interrupt controller (PIC), and bus accesses.

Once these build process defining configuration files are generated, cross compilation and linking produces the final target binary for the processor. This process uses a cross compiler specific to the target processor. It compiles the generated code (i.e. from code generation and HdS generation), as well the selected SW database components, into a binary suitable for the target processor.

The produced binary is then ready for execution on the target processor. The binary may be executed on the actual processor, for example on an FPGA-prototyping board, if the target hardware implementation description is available. Alternatively, for early evaluation if the target hardware platform is not available, the binary may be executed on an ISS-based virtual platform. An ISS-based virtual platform can be a TLM, as described in Chapter 3, in which the abstract processor model is replaced with an instance of an ISS. The ISS interprets the target binary of the generated software. The SW executed within the ISS has access to all registers and memory mapped I/O components equivalent to the hardware platform. From a software point, such a virtual platform already provides a very close match to an execution on the final hardware. Hence, an ISS-based virtual platform allows early validation of the final target binary.

This concludes the overview of software synthesis flow outlining the path from an abstract input model to execution of the final target binary. Before going into detail for each step, we will now introduce a possible input model, which we will use throughout the chapter.

5.2.1   EXAMPLE INPUT TLM

Figure 5.3 shows a sample system TLM as a possible input to the software synthesis. The TLM can be generated by system synthesis as we have described in Chapter 4. Alternatively, the TLM can be manually developed following guidelines for synthesizable semantics and features as we have described in Chapter 3. The approach for software synthesis and the amount of independent decision making depends on the abstraction level present in input

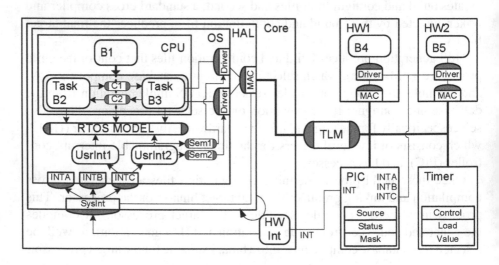

FIGURE 5.3 Input system TLM example

system TLM. As described in previous chapters, abstract models with varying degree of implementation detail are possible. When using a very abstract input model, which captures few implementation details, many implementation decisions need to be made during the software synthesis itself. On the other hand, a less abstract model, which reflects more implementation details, on the other hand, is better suited for software synthesis. To simplify software synthesis, the system TLM should reflect system decisions. These are decisions that affect the compositions of platform components and their interaction. In case, the system TLM reflects all system decisions, software synthesis can extract those decisions and generate a software implementation for them. Our example TLM in Figure 5.3 is a more detailed abstract model, that captures many system decisions.

The system TLM in Figure 5.3 contains two hardware units (*HW1* and *HW2*) and a processor, which is accompanied by a *Timer* and a *PIC*. All components are connected to the processor bus. Three modules are mapped to the processor: *B1*, *B2* and *B3*. Two of these modules are wrapped in tasks, *TaskB2* and *TaskB3*, as they execute concurrently. The tasks communicate with each other through channels *C1* and *C2*. In addition, both task communicate externally. *TaskB2* communicates with *B5* mapped to *HW2*, while *TaskB3* communicates with *B4* mapped to *HW1*.

A set of half-channels models the communication between the tasks on the processor and the modules mapped to the hardware units. We use the term "half channel" to indicate that a callable interface is only provided on one side of the channel. The other side is not callable and instead may by itself call another channel.half channel On the processor side, communication occurs through the half channels *Driver* and *MAC*. The latter connects to the processor bus. Matching half-channels are inside the hardware components *HW1* and *HW2*. In this example, both hardware units share the same interrupt, *INTC*. Both their interrupt lines connect to connected to *INTC* at *PIC*. The *PIC*, in turn, connects to the processors interrupt input, *INT*. The processor model contains an interrupt chain, which connects the processor interrupt to the appropriate driver. In the shown example, one interrupt chain starts with *SysInt*, followed by *INTC*, *UsrInt2*, and finally connects with *Sem2* via *Driver*.

The processor TLM is constructed in layers. It starts on the outside with the *Core* layer, followed by *HAL* and *OS*, with *CPU* as the innermost layer. Using this layering scheme as a classification, software synthesis produces an implementation for everything inside the *HAL* when creating a binary for the target processor.

To derive the embedded software from the TLM, software synthesis has to implement all SLDL language elements used inside the HAL (e.g. modules, tasks, channels and port mappings) on the target processor. The TLM is captured in an SLDL, and predominant SLDLs (e.g. SystemC, SpecC) are C or C++ extensions. Therefore, one possibility is to compile the selected portion of the TLM directly into binary code for the target microprocessor. However, such a direct compilation would produce a highly inefficient implementation. The microprocessor's basic SW would need to support the execution semantics of the SLDL, and therefore a large simulation kernel for the SLDL would be included in the compiled code, which may not be feasible considering the embedded system's resource constraints.

The reason for the large simulation kernel is rooted in the complexity of an SLDL. An SLDL is mainly geared toward modeling and simulation of designs at the system level. In order for them to be generally applicable and to handle a wide range of system architectures, much overhead is introduced to support system level features (such as hierarchy, concurrency, communication). This much overhead might be affordable when executing on the simulation host. However, not all features expressible in the SLDL are necessarily needed for the target software code. Considering the limited memory space and execution power of embedded processors, a direct compilation of the SLDL to the target micro processor is not suitable. Instead, software synthesis has to generate compact and efficient software code for implementation.

5.2.2   TARGET ARCHITECTURE

In the example input TLM above, we focus on a single processor system for ease of explanation. In more general, however, software synthesis targets a multi-core platform, as outlined in Figure 5.4. Such a platform may contain many processing elements, such as standard processors (e.g. *Proc 1*, *Proc N*) or hardware accelerators (e.g. *HW1*). Each processor may also contain local memory where its code is stored. Each processor may have a PIC, which allows the processor to listen to many incoming interrupt sources. In addition, each processor may have a local timer to perform time-related tasks, such as periodic execution or keeping track of time outs. A platform's processor is connected to the system through its processor bus. This allows communication with local

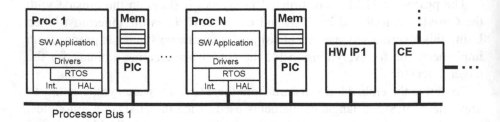

FIGURE 5.4 Generic target architecture

hardware components. Also, processors on the same bus can communicate with each other through global memory. More complex communication topologies are constructed with a Communication Element (*CE*), which connects one or more buses.

Using the decisions contained in the TLM, an automatic software synthesis generates the code for each processor. As outlined before, this involves generating the task code, support code for multi-tasking and internal communication, as well as driver code for external communication. The driver code implements communication with external hardware accelerators, external memory, and with the communication elements. In short, software synthesis generates all code to execute the applications distributed over the multi-core architecture.

The following sections focus on the synthesis for a single processor. The same procedure would be repeated for each processor. To allow communication between PEs, it is essential that the communication code inside communicating PEs implements matching system- wide decisions, as, for example, PEs would need to agree in the addresses they use. To achieve matching implementations, it is most beneficial to capture system-wide decisions already in the input system model (e.g. as decided by system synthesis). Then, software synthesis (as well as hardware synthesis) can generate matching implementations for each component and the complete multi-core embedded system can be constructed.

The next sections describe the synthesis process bottom up in detail. In Section 5.3, we start with code generation, which produces code for each task on the processor. Then, Section 5.4 outlines how multi-task synthesis creates code for the concurrent execution of these tasks. Section 5.5 shows how communication synthesis creates the drivers for internal communication while external communication is shown in Section 5.6. Finally, Section 5.8 describes combining all generated code to the target SW image.

5.3 CODE GENERATION

Code generation is the first step of software synthesis. It generates sequential code in the target language for each task within a programmable component. To produce the sequential task code, code generation uses the model of the application that is captured within the system TLM. The system TLM contains a representation of the application, consisting of a module hierarchy and a set of channels. The module hierarchy captures the application behavior. Modules declare communication interfaces with ports and channels connected to these channels express the communication. Hence, the system TLM contains a translation of the application originally described in the input MoC for convenient analysis, development, and synthesis. In the software synthesis stage, code generation then translates the application module hierarchy in the TLM into the target language for programming the processor. For the examples that follow in this chapter, we chose C as a target language.

Code generation translates the application module hierarchy into the target programming language. The TLM's application modules use system-level features of the SLDL, such as hierarchy, concurrency, and communication encapsulation, which are not natively present in target language C. Code generation must construct these SLDL features out of the available language constructs in order to implement them on the target processor. For example, it translates the hierarchical composition of modules in the SLDL into flat C-code containing functions and data structures. Attention is needed for module local variables, as ANSI-C does not provide such an encapsulation. A module's local variables can be added to a module-representing structure. Then, for each module instance, an instance of a particular structure is created. Communication between modules need to be addressed too. For modules within the same task, their communication can be represented as function arguments. Modules in different tasks can communicate via inter-process communication. On top of these translations, SLDL specific extensions, such as bit vectors and events, have to be implemented on the target.

The main idea of the conversion process from SLDL to ANSI-C is to convert a module or channel into a C `struct` and a set of C functions. The module hierarchy can then be translated into a C `struct` hierarchy. In some ways, this translation process is similar to the one by early C++ to C compilers, when translating a C++ class hierarchy to flat C code. We now present simplified rules for code generation's conversion process. These rules apply equally to modules and channels. To facilitate a more straight-forward explanation, we will focus on the modules. The rules for C code generation are as follows:

Rule 1: Each module is converted into a C `struct`.

Rule 2: The structural hierarchy among modules is represented in a C `struct` hierarchy. Child modules are instantiated as `struct` members inside the parent `struct`.

Rule 3: Variables that are defined inside a module are converted into data members of the module representing C `struct`.

Rule 4: Ports of a module are converted into data members of the module representing C `struct`.

Rule 5: Methods inside a module are converted into global functions. An additional parameter that represents the module instance to which the function belongs is added to each global function.

Rule 6: A static `struct` instantiation for the whole processing element is added at the end of the output C code. It contains the `struct`s of

```
1  SC_MODULE(B1){
2    int A;
3    sc_port<iChannel> myCh;
4    SC_CTOR(B1){}
5    void main(void) {
6      A = 1;
7      myCh->chCall(A*2);
8    }
9  };
10
11 SC_MODULE(TaskB2){
12   CH1 ch11, ch12;
13   B1 b11, b12;
14   SC_CTOR(TaskB2):
15     ch11("ch11"), ch12("ch12"),
16     b11("b11"), b12("b12") {
17     b11.myCh(ch11); // connect ch11
18     b12.myCh(ch12); // connect ch12
19   }
20   void main(void) {
21     b11.main();
22     b12.main();
23   }
24 };
```

LISTING 5.1 SystemC task specification

all converted modules. Thus, it allocates the data used by the PE's software. Port mappings for modules and channels inside the task are established in this `struct` initialization.

A simple example will help us illustrate the application of these rules and to explain the code-generation process. Figure 5.5 depicts a module hierarchy for conversion and Listing 5.1 shows the corresponding SystemC code. Listing 5.2 outlines the output ANSI-C code.

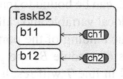

FIGURE 5.5 Task specification

In the example shown in Figure 5.5, *TaskB2* consists of two instances of module *B1*, namely *b11* and *b12*, that execute sequentially. Each module instance is connected to its own channel instance of type *CH1*. So, module instance *b11*

```
 1  struct B1 {
 2     struct CH1 *myCh; /* port iChannel*/
 3     int a;
 4  };
 5  struct TaskB2 {
 6     struct B1 b11, b12;
 7     struct CH1 ch11, ch12;
 8  };
 9  void B1_main(struct B1 *This) {
10     (This->a) = 1;
11     CH1_chCall(This->myCh, (This->a)*2);
12  }
13  void TaskB2_main(struct TaskB2 *This){
14     B1_main(&(This->b11));
15     B1_main(&(This->b12));
16  }
17  struct TaskB2 taskB2= {
18     {&(taskB2.ch11),0/*a*/}/*b11*/,
19     {&(taskB2.ch12),0/*a*/}/*b12*/,
20     {} /*ch11*/, {} /*ch12*/
21  };
22  void TaskB2() {
23     TaskB2_main( &task1);
24  }
```
LISTING 5.2   ANSI-C task code

connects to channel instance *ch11* and *b12* to *ch12*. For brevity, the example does not show other modules within the task and also omits communication outside the task.

Listing 5.1 outlines same example as an SystemC specification containing two modules *B1* and *TaskB2*. *B1* is defined in lines 1-9 starting with SC_MODULE(B1). The module contains a local variable *A* (line 2) and a port *myCh* (line 3). The port indicates that module *B1* requires an interface and that it will call methods of that interface. Later, a channel or module, which provides the required interface, can be bound to that port. In the *main()* method of module *B1*, it accesses the local variable and calls the method *chCall()* on its port (line 7). Depending on the binding of the port, a channel implementation will be called. Both instances of module *B1* are port bound to an instance of *CH1*. Therefore, the port call in line 7 will result in the execution of method *chCall()* in one channel instance of *CH1*.

The declaration of *TaskB2* extends from line 11 through 24, beginning with SC_MODULE(TaskB2). It contains two instances of *B1*, namely *b11* and *b12*. Line 13 shows their instantiation. Line 16 defines the names of the child modules with b11("b11"), b12("b12"). *TaskB2* also contains two instances of channel

CH1, namely *ch11* and *ch12* (line 12). The constructor of *TaskB2* (lines 17 and 18) connects the channel instances to the ports of *b11* and *b12*. For example, b11.myCh(ch11), line 17, connects *ch11* to the port of module instance *b11*. In its method *main()* starting with line 20, *TaskB2* sequentially calls the main methods of *b11* and *b12*.

Listing 5.2 shows the output ANSI-C code. It defines a C struct for each module, a global function for each module method, and instantiates a global struct for the module data and port mapping. We can find examples of the applied code generation rules within the listings:

Rule 1: Module *B1* is converted into struct *B1* (lines 1-4). Module *TaskB2* is converted into struct *TaskB2* (lines 5-9).

Rule 2: In the input code, module *TaskB2* contains two instances *b11* and *b12* of module *B1* (line 13 in Listing 5.1). Correspondingly, the struct *TaskB2* contains two instances of struct *B1* with the names *b11* and *b12*, as shown in line 6 of the output C code.

Rule 3: The module local variable, int *A*, defined in module *B1* (line 2 in the input) is converted to an identical data member inside struct *B1*. See line 3 in Listing 5.2.

Rule 4: The port of module *B1* (line 3 in Listing 5.1) is represented by a pointer to the connected channel inside the struct *B1* (line 2, Listing 5.2)

Rule 5: The method *main* inside module *B1* is converted to a global function B1_main() in the output C code (line 9). One additional parameter (struct B1 *this) is added referencing the context. This parameter is needed to distinguish between different instances of the same module, as, for example, module local variables may have distinct values in different instances. To distinguish between these instances, the generated global function (B1_main()) is called with a pointer to the module representing struct instance. See the calls in lines 14, 15. They differ in the argument that passes the context. One refers to the instance *b11* (This->b11), and the other to *b12* (This->b12).

The code inside *B1_main()*, which accesses data member of *B1*, is converted to use references to data members inside the struct *B1*. For example, inside function B1_main(), the variable *A* in the input code is now used as (This→A) in the output code (line 8 in Listing 5.2). As a result, each instance of *B1* in the output code retains its own copy of the local variables. For example, the local variable *A* is initialized with 0 for both instances.

Rule 6: The data used by task *TaskB2* is statically allocated through the instantiation of the top level struct *TaskB2* (see lines 17 to line 21 in Listing 5.2). The initial values for data members inside struct *TaskB2* are all set at this time.

The port mapping information is also recorded at this point. In our example, module *B1* contains one port. The mapping for the instance *b11* is set in line 18 in Listing 5.1. It refers to channel instance *ch11*, which is also a member of struct *TaskB2*. This approach in implementing port mapping has the advantage of being set at compile time rather than at runtime. It therefore reduces the runtime overhead to a minimum. Other, more dynamic approaches may result in a higher runtime overhead.

Please note that our example implements an optimization for calls to channel methods. In our case, the call to the method *chCall()* (line 7, Listing 5.1) is translated directly to a function call with a context pointer (line 11, Listing 5.2). This optimization is possible because all instances of *B1* map the port *myCh* to instances of the same channel, namely *CH1*. Therefore, the method implementation is identical, and the global function *CH1_chCall()* representing the channel method can be called directly. In a more general case, two instances of the module may map to different channels which both implement the same interface. Separate global functions would then need to be called. In that case, a significantly more elaborate solution involving virtual function tables would be necessary. We are omitting this case here for brevity, but a detailed description more general translation can be found in [197].

The outlined procedure assumes that the code inside a module is C compliant, with the exception of calls to ports. This simplifies the code generation to resolve SLDL specific-features, as described above. A significantly more complex approach would be required if the full C++ language feature set were to be supported inside module methods. Then, for example, C++ libraries would have to be re-implemented or converted into C, which could render the solution infeasible in terms of effort and memory footprint.

Language complexity is the main challenge to the solution outlined above. It requires a parser for the SLDL that can extract module hierarchy and connectivity. SystemC being a library extension of C++ allows a very flexible model construction, which may complicate parsing. For example, static and dynamic object allocation (via new()) have to be detected, and object accesses via pointer or value has to be supported. Port connectivity, to give a further example, is typically captured in the constructor. However, mapping can also occur in any hierarchy of method calls as long as mapping is completed at the end of the elaboration phase. Supporting this freedom can make efficient model parsing infeasible. In order to enable an efficient synthesis process, only a subset of SLDL features can be allowed and strict model guidelines are required.

5.4 MULTI-TASK SYNTHESIS

When multiple tasks are mapped to the same processor, they have to be scheduled to alternate their execution. Multi-task generation produces code that uses an underlying multi-task engine in order to manage and schedule such tasks.

The following pages focus on two possible approaches for dynamic multi-tasking: RTOS-based multi-tasking and interrupt-based multi-tasking. The approach predominantly-used is RTOS-based multi-tasking, in which user tasks are executed on top of an off-the-shelf RTOS and scheduled by the RTOS scheduler. Sometimes performance and resource constraints hinder using a complete RTOS. In such a case, an alternative of interrupt-based multi-tasking can be applied. Here the generated application executes on a bare processor using interrupts and does not require any operating system. Interrupt-based multi-tasking, however, is only suitable for systems with few tasks. Resource constraints permitting, an RTOS-based solution is preferred for its flexibility,

5.4.1 RTOS-BASED MULTI-TASKING

As we introduced in Section 5.1.2, embedded systems frequently use an RTOS for dynamic scheduling of tasks. We call this "RTOS-based multi-tasking." For this, off-the-shelf RTOSes are popular with developers because they typically are reliable, well-tested operating systems that offers great flexibility. In addition they often come with significant tool support from the RTOS vendor. Often, they are highly configurable to tailor the OS to the application needs. Through configuration, the memory footprint can be minimized to fit the needs of the embedded system being designed.

Figure 5.6 shows a generic software stack for RTOS-based multi-tasking. The stack constists of HAL, interrupts, RTOS, RAL, and application.

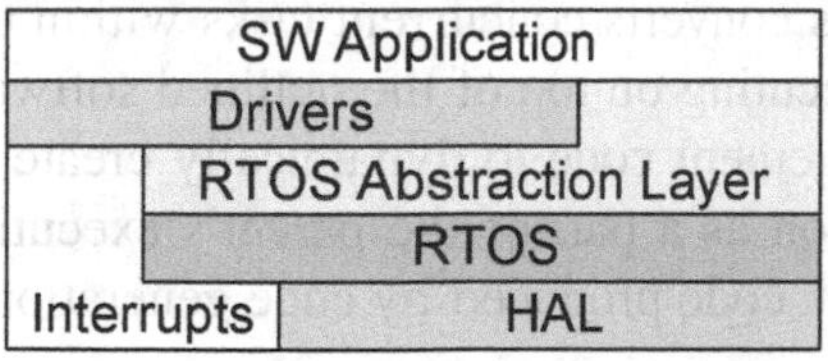

SW Application	
Drivers	
RTOS Abstraction Layer	
RTOS	
Interrupts	HAL

FIGURE 5.6 Software execution stack for RTOS-based multi-tasking

At the bottom, the *HAL* abstracts the physical hardware of the processor from the software that is running on top. It hides differences in hardware programming so that the operating-system code can be mostly independent from the underlying hardware. For example, the *HAL* provides facilities for saving and restoring the processor's internal state. The operating system uses these facilities to switch between tasks. The HAL implements low-level drivers for communication on the processor bus. It provides a communication interface with the PIC for registering interrupt handlers and for evaluating the interrupt status. It also supplies facilities to program timer module.

At the same level as the *HAL*, *Interrupts* are used for synchronization with external devices. Above *HAL* and *Interrupts*, the *RTOS* provides services for task management, communication, timing management.

On top of the *RTOS*, an RTOS Abstraction Layer (*RAL*) can be used in order to provide a canonical OS interface. It abstracts from a particular OS's function names and parameters. As a result, the canonical OS interface limits inter-dependency between synthesis and the actual target RTOS. This significantly reduces effort for customizations within the synthesis flow when supporting a wide range of RTOS implementations. RTOS-implementations may differ in the API they use. Standardized APIs (e.g. POSIX, OSEK, ITRON) exist, which target specific application domains. In addition, many RTOS-specific and proprietary APIs are in use (uCOS-II, vxWorks, eCos, RTEMS). To ensure a sufficiently generic RAL, many RTOS APIs have to be investigated so that common primitives for task scheduling, communication and synchronization can be chosen. These become the basis for multi-task synthesis. Typically, RTOSes provide a very similar set of basic primitives, such as task creation, semaphores, and timing delay. Wrapping them results in a very thin abstraction layer. In the case that a required primitive should not be available in a particular RTOS, an emulation has to be constructed out of available primitives.

One layer above the *RAL*, *Drivers* implement application-specific communication with external components, using services from *RAL* (e.g. for internal communication), *HAL* (bus access for communication), as well as *Interrupts* (for synchronization with external components). Finally, the *SW Application* executes on top of the stack. It directly uses only communication *Drivers* and services of the *RAL*.

Multi-task synthesis converts concurrent tasks within the system TLM into RTOS-based tasks executing on top of the outlined software stack. It involves generating task-management code to dynamically create tasks and then waiting for their termination as a part of the parent's execution. Each task itself uses the sequential task code produced by code generation (as we explained in Section 5.3).

Figure 5.7 depicts a portion of a system TLM as an example for RTOS-based multi-task synthesis. The same example is shown in Listing 5.3 in its SystemC specification. Finally, Listing 5.4 outlines the output code.

The input model shows the mapping of modules to tasks. The example in Figure 5.7 contains two parallel executing tasks *TaskB2* and *TaskB3*. They execute within the module *B2B3*.

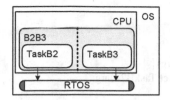

FIGURE 5.7 Multi-task example model

Listing 5.3 outlines the SystemC definition of module *B2B3*. It instantiates the child tasks in lines 4 and 5. In the constructor, lines 6 - 11, the task parameters are defined. Parameters contain task name, priority, and stack size. The *main()* method of *B2B3* starts in line 12. It releases first the child tasks using the *release)(* function. As a result, *TaskB2* and *TaskB3* start executing concurrently to *B2B3*. Consequently, *B2B3* then waits until each task finishes using the *join()* function.

Listing 5.4 shows an excerpt of the output ANSI-C code implementing this example on the target. The basic outline follows the principles for code generation explained in Section 5.3. It contains global functions and structures representing the modules. During synthesis, the task-control information is extracted from the TLM and the synthesis generates task management calls using the RAL API. The release statements in the TLM (line 12, 13 in Listing 5.3) are replaced with *taskCreate()* calls, which dynamically create and release tasks;

```
1  SC_MODULE(B2B3) {
2  public:
3    sc_port<iRTOS> rtos;
4    TaskB2 taskB2;
5    TaskB3 taskB3;
6    SC_CTOR(B2B3):
7      taskB2("taskB2", 5, 4096),
8      taskB3("taskB3", 2, 4096) {
9      taskB2.rtos(rtos);
10     taskB3.rtos(rtos);
11   }
12   void main(void) {
13     taskB2.release();
14     taskB3.release();
15     taskB2.join();
16     taskB3.join();
17   }
18 };
```

LISTING 5.3 Multi-task example SystemC code

```
1  struct B2B3{
2      struct TaskB2 task_b2;
3      struct TaskB3 task_b3;};
4  void *TaskB2_main(void *arg){
5      struct TaskB2 *this=(struct TaskB2*)arg;
6      /* ... */
7  }
8  void *TaskB3_main(void *arg){
9      struct TaskB3 *this=(struct TaskB3*)arg;
10     /* ... */
11 }
12 void *B2B3_main(void *arg){
13     struct B2B3 *this= (struct B2B3*)arg;
14     os_task_handle task_b2, task_b3;
15     task_b2 = taskCreate(TaskB2_main,
16                        &this->taskB2, 5, 4096);
17     task_b3 = taskCreate(TaskB3_main,
18                        &this->taskB3, 2, 4096);
19
20     taskJoin(task_b2);
21     taskJoin(task_b3);
22 }
```

LISTING 5.4   Multi-task example ANSI-C code

see lines 17-20 in Listing 5.4. For task creation, the task's parameters, such as priority and stack size, are extracted (lines 4, 5 in Listing 5.3) and passed as arguments to *taskCreate()*. The task then executes the flattened C code as produced by the code generation explained before. The implementation of the task management itself is hidden inside the RAL. After task creation, *B2B3_main()* waits for the completion of the created tasks by using the *taskJoin()* function in lines 20 and 21.

The output code shown above uses the RAL services independently from the actual RTOS. During the synthesis, the designer can select a suitable RTOS. Later, in binary image creation, which we will describe in Section 5.8, the selected RTOS, together with a specialized abstraction layer, will be included in the final binary.

5.4.2   INTERRUPT-BASED MULTI-TASKING

Interrupt-based multi-tasking is an alternative option for dynamic scheduling. For some specific processing elements, an execution on top of an RTOS may not be desirable. This may be the case when the processing element consist of very few tasks, when the code is targeted to execute on a DSP, or when strict memory footprint limitations rule out utilizing an RTOS. In such cases, interrupt-based multi-tasking can target a bare processor, on which concurrent software execution is performed without any RTOS. Instead, interrupts are utilized to provide multiple flows of execution.

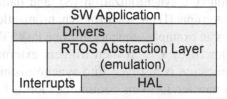

FIGURE 5.8 Software execution stack for interrupt-based multi-tasking

Figure 5.8 shows the software execution stack for interrupt-based multi-tasking. The stack is almost identical to the RTOS-based stack shown earlier in Figure 5.6, the difference being that here the RTOS is missing. The *RAL* is larger as it implements a partial RTOS emulation, but this emulation is very thin because it provides only a fraction of the RTOS services (e.g. simple events, processor suspension).

In this scenario, the RAL does not provide task-management code. Instead, the code inside a task has to be specially generated for interrupt-based multi-tasking. To give an intuitive explanation, we can say that multi-task synthesis converts the lowest-priority task to execute in the processor's main function. All other tasks are converted into state machines, which then execute in the context of interrupt handlers.

Interrupt-based multi-tasking therefore relies heavily on state machines. To understand why states are needed, contrast the interrupt-based against the RTOS-based solution. Each task running on top of an RTOS has an own stack. In a preemptive multi-tasking the RTOS can alternate between tasks at any point within the task execution. To switch between tasks, it first stores the status and context of the current task on the task's stack. The RTOS then restores the status of the new task and continues with the new task's execution.

In interrupt-based multi-tasking, on the other hand, tasks share the same stack. Hence, we cannot use the stack to store the task's context when switching. Instead, we break a task into individual states, which are executed in a state machine. Each task gets its own state machine. After the completion of each state, we can switch between the state machines of two different tasks. The context of a task, i.e. where it should continue, can be reduced to the current state of its state machine. So in interrupt-based multi-tasking, we use states to alternate between tasks and to minimize the task context data to be kept while switching.

The next paragraphs describe in detail the process of converting task code for an application into a state machine for interrupt-based concurrent execution.

An application task is composed of application modules, which capture computation, and calls to communication drivers. The driver code communicating with external hardware contains both synchronization and communication. Therefore we can more formally assume that each task is composed of a sequence of computation (C), synchronization (S), and data transfers (T). We further assume that interrupts (I) are used for synchronization.

Figure 5.9(a) shows an example sequence for one task. The tasks starts with computation *C0*, followed by *C1*. It then triggers external communication. The external communication is resolved into synchronization *S1*, which uses interrupt *I1*, and data transfer *T1*. This communication is followed by another set of computation *C2* and communication (consisting of *S2*, which uses *I2*, and *T2*). Following that, execution loops back to *C1*.

If only interrupts are used for synchronization (*S1* and *S2*), then the task's main function can be split into a state machine. A new state is created each time a synchronization point (S), a loop, or a conditional execution is encountered in the generation process.

Figure 5.9(b) shows the output state machine consisting of four states. State *ST1* has been created because *C1* is the first element inside a loop. The inserted distinction between the states *ST0* and *ST1* accommodates the one-time execution of *C0*, while *C1* is repeated in the loop. State *ST2* was created due to synchronization *S1*, while *ST3* was created due to synchronization *S2*. During execution, the state machine transitions to the next state upon successful synchronization. In the example, *S1* uses interrupt *I1*. Upon receiving interrupt *I1*,

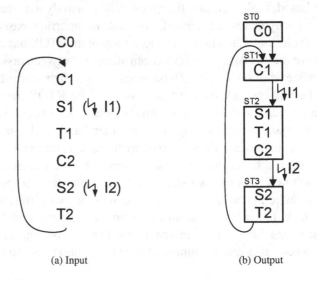

(a) Input   (b) Output

FIGURE 5.9   Interrupt-based multi-tasking example

the state machine transitions from *ST1* to *ST2*. By converting a task into states, we can switch between tasks at the boundary of states without needing the stack support of an RTOS-based solution. For example, if synchronization *S1* has not occurred after finishing state *ST1*, we can switch to executing another task's state machine. Then, when *S1* does occur, the original state machine can resume with *ST2*. We will later offer a code example that highlights that sequence.

Splitting the original task into individual states, however, demands special attention to local variables. Local variables can still be used within a state. However, they can not be used for storing data across states, since the state machine may terminate and a different task's state machine may be executed. Hence, these local variables may lose their content when resuming the state machine. Therefore, all variables that are used across states have to be moved into a global data structure. In a simple approach, each local variable of a task's main function can be integrated into a task-specific global data structure. The created task's state machine is then executed in the interrupt handlers, which are originally used for synchronization. In the above example, the state machine is executed in the handlers of *I1* and *I2*. The generated task code executes incrementally in separate invocations of the interrupt handler.

To summarize the conversion, let's look again at our goal: the task has to be converted so that it no longer relies on an own stack when switching between tasks. We achieve this by converting the task into a state machine and by: (a) allowing switches between tasks only at the state boundary, (b) capturing the execution progress within a task in its state machine status, and (c) moving local variables that are carried between states are into a task-specific global data structure. As a result, a task converted to a state machine can execute incrementally in interrupt handlers.

This conversion process also allows for the preservation of task priorities for the converted tasks. This, however, depends on the priority distribution of the interrupts selected for synchronization. Interrupts (with their respective priority) have to be selected according to task priority. Assume multiple bands of interrupt priorities: to preserve the task's priority, a higher-priority task has to use interrupts out of the higher-priority band, exclusively. Conversely, a lower-priority task must use interrupts only from the lower-priority band. The lowest priority task on the processor can execute in the processor's startup task (with executes *main()*. As a result, the prioritized execution of interrupts preserves the task priorities.

We now describe an example implementation of a task state machine. Listing 5.5 outlines the C implementation for the state machine introduced in Figure 5.9(b). The excerpt shows the interrupt handler in function *intHandler_I1()* and the task state machine in *executeTask0()*. The interrupt handler implements synchronization S1 and executes the task state machine. The state machine is implemented in *exectueTask0()* with a do-while loop containing a switch-case

```
1  /* interrupt handler */
2  void intHandler_I1() {
3     release(S1);   /* set S1 ready */
4     executeTask0(); /* task state machine */
5  }
6  /* task state machine */
7  void executeTask0() {
8     do { switch(Task0.State) {
9        /* ... */
10       case ST1: C1(...);
11             Task0.State = ST2;
12       case ST2: if(attempt(S1)) T1_receive(...);
13             else break;
14             C2(...);
15             Task0.State = ST3;
16       case ST3: /* ... */
17    } } while (Task0.State == ST1);
18 }
```

LISTING 5.5 State machine implementation

construct. The current state is captured in a global variable *Task0.State*. Assume for this explanation that the task's state machine is currently executing in the interrupt handler for *I1* and the current state is *ST1*.

After finishing the execution of *C1* in line 10, the new state is set at *ST2*. At the beginning of the new state, the synchronization *S1* is checked with *attempt(S1)*, line 11. In case the synchronization has not yet occurred, the state machine terminates with the `break` statement (line 14). Consequently, the do-while loop, the function *executeTask0()*, as well as the interrupt handler, all terminate so that the processor can serve a lower-priority interrupt or the main function.

Upon receiving the next interrupt, *I1*, the registered interrupt handler *intHandler_I1()* (line 1) is executed. In line 2, the handler signals that *S1* is ready and then calls the state machine again (line 3). The current state is still *ST2*, therefore the condition in line 11 is tested again. The test *attempt(S1)* now passes, since the synchronization has occurred. The task continues by receiving the data (line 12) and subsequently executing the computation *C2* in line 16. The switch-case statement (lines 7 to 20) is surrounded by a do-while loop, which is required to implement loops between states. In this example, the loop is necessary to transition from state *ST3* back to ST1 without terminating the interrupt handler.

The presented approach for interrupt-based multi-tasking is an efficient alternative for those occasions when design constraints disallow the use of an RTOS. It is best targeted to systems with very few tasks, which permits us to efficiently convert those tasks into state machines. One drawback of this state machine conversion is that the task code may suffer in readability, since it is split across states. Another consideration when using interrupt-based multi-tasking is that careful planning is required for the interrupt mapping and priority distribution to maintain the system's responsiveness, as the application would be mainly executing in interrupt handlers.

5.5 INTERNAL COMMUNICATION

Internal communication, or Inter Process Communication (IPC), takes place between tasks on the same processor. In the example Figure 5.10, the channels *C1*, *C2*, *Sem1*, and *Sem2* are used for internal communication. These are instances of standard channels that are supported by the design flow.

To realize a particular communication on the target system, the abstract standard channels in the simulation model are replaced with a target-specific implementation. The target-specific implementation then uses the primitives of an underlying RTOS, such as semaphores and events. Note that this implementation should not recreate the simulation environment on the target. A target specific implementation should instead recreate the same interface and same semantics as the abstract channels.

To give an example, Listing 5.6 shows a code excerpt implementing a single handshake channel for internal communication. A single handshake offers one-way synchronization with storage for one event. Listing 5.6 is specific to the Xilkernel, a Xilinx proprietary RTOS. It contains the definition of struct *tESE_ch_shs* (lines 2-4) for capturing the channel state and variables, function *ESE_shs_init()* for initializing the channel, as well as send and receive func-

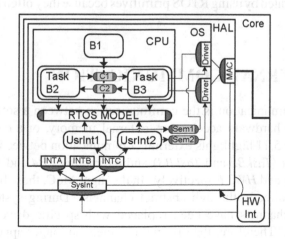

FIGURE 5.10 Internal communication

```
1  /** SHS OS—specific struct */
2  typedef struct {
3      sem_t req; /**< os semaphore */
4  } tESE_ch_shs;
5  void ESE_shs_init(tESE_ch_shs *pThis){
6      int retVal = sem_init(&pThis->req, 0, 0);
7      /* ... error handling */
8  }
9  void ESE_shs_send(tESE_ch_shs *pThis){
10     int retVal = sem_post(&pThis->req);
11     /* ... error handling */
12 }
13 void ESE_shs_receive(tESE_ch_shs *pThis){
14     int retVal = sem_wait(&pThis->req);
15     /* ... error handling */
16 }
```

LISTING 5.6 Internal communication example of single handshake

tions, *ESE_shs_send()* and *ESE_shs_receive()*. Due to the close match in semantics, the single handshake is implemented directly with an RTOS primitive: a semaphore. The send function directly calls *sem_post()* (line 10) and conversely receive calls *sem_wait()* (line 14). Note that, similar to the principle explained in Section 5.3, we use a data structure to maintain the status of the channel. In the example, *tESE_ch_shs* (lines 2 - 4) contains a single member (sem_t req;) referring to the OS semaphore. A pointer to the structure instance is passed as an argument upon calling a channel function.

As shown in the example above, internal communication channels are efficiently implemented by using RTOS primitives because they often match closely in semantics.

5.6   EXTERNAL COMMUNICATION

External communication is the communication between a software process and an external hardware accelerator, external memory, or a communication element. Figure 5.11 highlights external communication between the two tasks on the processor (*TaskB2* and *TaskB3*) and the modules *B4* and *B5*, which are mapped to *HW1* and *HW2* respectively. In the input MoC, these behaviors have communicated directly through abstract channels. During system synthesis, these abstract channels have been replaced with specific drivers for system communication. Therefore, the original channels no longer appear directly in the system TLM. Instead, the original abstract channels are resolved into stacks of half channels (namely *Driver* and *MAC*). These half channels are inserted into the processor model. A matching stack of half channels is present in each HW component (*HW1* and *HW2*). These matching stacks implement the same system-wide communication decisions and therefore enable communication.

To support heterogeneous systems, we can follow concepts from the ISO/OSI layering model [98] as we implement external communication. To implement this communication decisions about various aspects are needed: the network byte layout, a selection of channels to merge, the packet size, packet switching and routing. We discussed these communication decisions in Chapter 4. For synthesis it beneficial, if the communication drivers inside the system TLM contain this information The following sections detail how HdS synthesis implements these decisions on a software-processing element. Note that in the interests of brevity, our description will focus on the communication with a synthesized hardware component. Using a synthesized component guarantees matching communication stacks and procedures. Incorporating IP components of other manufactures may require additional effort if the communication stacks are not matching. Such a case would require either a synthesized hardware wrapper, which converts the IP protocol to the system protocol, or a specialized software driver, which produces the IP protocol.

5.6.1 DATA FORMATTING

marshalling

Communication between heterogeneous processing elements involves unique challenges. Two communicating processing elements may have different memory layouts, due, for example, to different byte orders, also referred

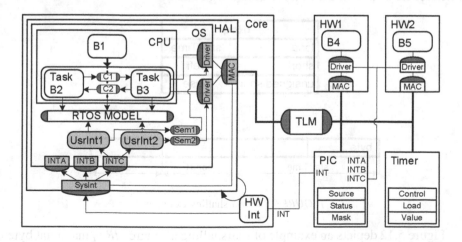

FIGURE 5.11 External communication

```
1  typedef struct stReq {
2    long          startTime;
3    short         coeff1;
4    unsigned short base;
5  } tReq;
```
LISTING 5.7 User type definition

as endianess. In that case, identical data would appear differently in memory depending on which processing element writes the data. To communicate between two such heterogeneous processing elements, common data-formatting rules have to be used between communication partners. This data formatting applies both for messages transferred over the communication media as well as for variables stored in common memory. In the simplest cases, both communicating processing elements natively have identical data-formatting rules (for example in terms of byte order, bit widths, padding, and packing rules), so no translation is necessary. If, however, the data formats differ, then the data has to be converted between the processor native layout and a common network layout.

The process of converting data from the processor native layout to the network layout is called "marshalling." Marshalling converts the user data into a flat, untyped data stream. Any processing element can interpret this data stream by using the information about the network data layout. The reverse process, converting from the network layout to processor native layout, is called "demarshalling." Marshalling data is a common issue in heterogeneous system communication. The CORBA standard [148], for example, defines elaborate rules for its Common Data Representation (CDR).

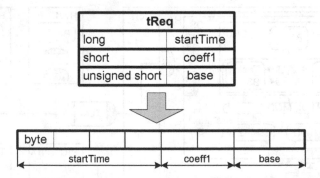

FIGURE 5.12 Marshalling example

Figure 5.12 depicts an example of marshalling a `struct` *tReq* into a flat byte stream. Listing 5.7 shows the corresponding data-structure definition. It shows

```
1  void myCh_send(/* ...*/ *This, struct tReq *pD){
2    unsigned char *pB = This->buf;
3    htonlong(pB, pD->startTime);
4    pB += 4;
5    htonshort(pB, pD->coeff1);
6    pB += 2;
7    htonushort(pB, pD->base);
8    pB += 2;
9    DLink0_trans_send(/*...*/This->buf, 8);
10 }
```

LISTING 5.8 Marshalling code

struct *tReq*, containing three elements *startTime*, *coeff1*, and *base*. In the generation process, the user data type definition has to be extracted from the system model and application-specific marshalling code has to be generated, which then serializes the user-specific structure data into the flat byte stream. Listing 5.8 shows an example of data marshalling. The code iterates through each struct member. It uses standard a conversion function to convert the primitive data type into a flat stream. For example, line 3 shows converting a long). Marshalling constructs the untyped message incrementally. The message is finally passed to the next layer in line 9. Note that for explanation purposes, we show the marshalling code as a separate function. For a more efficient implementation, it may be inlined as well.

As indicated above, the marshalling code producing an untyped data stream is highly application specific. If the user code only uses primitive data types, marshalling is straight-forward. However, arbitrary complex code may be necessary if the user data is constructed hierarchically out of user-defined types and arrays.

5.6.2 PACKETIZATION

packetization

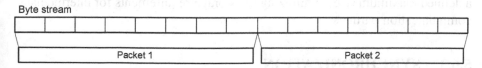

FIGURE 5.13 Packetization

Marshalling creates an untyped data stream. The length of this untyped data stream depends on the user data transferred and hence is arbitrary in length.

```
1  DLink0_trans_send(void *pMsg, unsigned int len){
2    unsigned char *pPos = pMsg;
3    while(len) {
4      unsigned long pktLen;
5      /* length is minimum of max size and len */
6      pktLen = min(len, CONFIG_PACKET_SIZE);
7
8      DLink0_net_send(pPos, pktLen); /* transfer */
9
10     len   -= pktLen; /* decr. transferred len */
11     pPos += pktLen; /* advance pointer */
12   }
13 }
```
LISTING 5.9 Packetization code example

However, only limited storage capability may exist in the communication partners along the route. To reduce the storage requirements, the untyped data stream can be split into smaller packets before transfer. On the receiving side, the packets will then be reassembled to the complete data stream before demarshalling. Options for packetization include fixed-sized packets, in which the unused portion of the packet is filled with padding bytes, and variable-sized packets. Variable sized packets necessitate the transmission of the packet size, unless all communication partners know the packet sizes a priori.

Figure 5.13 depicts how packetization breaks a large untyped byte stream into smaller packets. Listing 5.9 outlines the implementation of packetization. In a while loop (lines 3 through 12), the untyped message (*pMsg*) is split into smaller packets and transmitted through the network layer (line 8). The maximum packet size is captured in the constant CONFIG_PACKET_SIZE. The actual packet size is determined in line 6 by a minimum of the remaining length and the maximum size. After transmission, the remaining length (*len*) and the position to the next packet (*pPos*) are updated in line 10 and 11, respectively. The while loop terminates once all bytes in the data stream have been transferred. As result of packetization, the arbitrary-length input data stream is split into packets with a defined maximum size, minimizing the storage requirements for intermediate communication partners.

5.6.3   SYNCHRONIZATION

synchronization One further aspect of external communication is synchronization. Synchronization is required to signal that a communication partner on the same link is ready for a data transfer. Synchronization applies to both directions, for receiving, it signals that required data is available. During send-

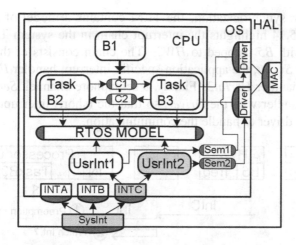

FIGURE 5.14 Chain for interrupt-based synchronization

ing, it ensures that the HW unit is ready for receiving data. Some bus protocols include synchronization semantics on the protocol level and demand that communication partners be always available. In such a case, synchronization may not be required at the link level in software. In the case of typical master/slave busses, however, synchronization is required. For these situations, the designer chooses the type of synchronization for each channel, selecting between polling or interrupt-based synchronization. Furthermore, the designer may choose in interrupt-based synchronization to share interrupts among sources to reduce the overall number of interrupt pins.

The following paragraphs describe the basic two forms of synchronization. In addition to strict interrupt and polling synchronization, hybrid forms and variants are possible. For example, a hybrid form of synchronization may alternate between variants based on heuristics, such as the fill status of a queue. The choice of the most suitable synchronization method depends on various characteristics, such as the duration between triggering a synchronization and testing for it, latency requirements, interrupt inter-arrival time, and tolerance for interrupt overhead.

INTERRUPT SYNCHRONIZATION

Interrupt-based synchronization uses a wire, additional to the processor bus, for sending an asynchronous signal from a hardware unit to the processor. Upon receiving of the interrupt, the processor saves the current state of execution, and starts executing an interrupt handler to react to the asynchronous signal. For example, the interrupt handler may initiate the communication with the hardware unit, or alternatively release a user task to perform this communication.

For interrupt synchronization, the TLM contains a model of the interrupt chain. Figure 5.14 highlights the interrupt chain in the system TLM for synchronization with *B5* mapped to *HW2*. The chain consists of the system interrupt handler *SysInt*, the application-specific interrupt handler *INTC*, and the user interrupt handler *UsrInt2*. Finally, a semaphore channel, *Sem2*, connects the interrupt handler with the *Driver*, so that the (short) interrupt handler can start the (long) driver to handle the communication.

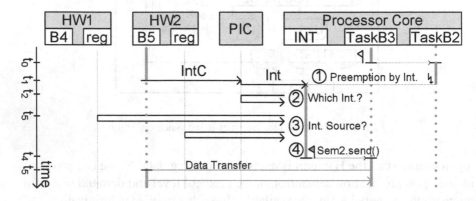

FIGURE 5.15 Events in interrupt-based synchronization

To implement interrupt-based synchronization, the HdS generation produces a chain of correlated code. The next paragraphs describe this interrupt-based synchronization code. The explanation follows an event sequence of sending a message from *B5*, which is mapped to hardware component *HW2*, to *TaskB3*, which is mapped to the processor. Figure 5.15 illustrates that event sequence

At t0, the *TaskB3* expects a message from *B5*. With the message not yet available, *TaskB3* waits on the semaphore *Sem1* and yields execution to the next lower priority task, *TaskB2*. At t1, behavior *B5*, which is mapped to *HW2*, reaches the code to send the expected message. Via interrupt *INTC*, it signals to the processor core the availability of the message . On the way, the *PIC* sets the processor interrupt *Int*. This in turn triggers the interrupt chain on the processor, which we have labeled with steps 1 through 4:

1 In step 1, the low-level assembly interrupt handler preempts the currently running task, *TaskB2*. It stores the current context on the stack and then calls the system interrupt handler. The low-level assembly interrupt handler is part of the RTOS port and is inserted from the software database.

2 In step 2, the system interrupt handler (half channel *SysInt*) communicates with the *PIC*. It determines through memory-mapped I/O the highest-priority pending interrupt. It then invokes the application-specific interrupt handler

(half channel *INTC* in the TLM). The *SysInt* code is one element of the Hardware Abstraction Layer (*HAL*) stored in the database.

3 In the example platform, the interrupt is shared between *HW1* and *HW2*, so the next step in the synchronization is to determine which of these is the source of the interrupt. The application-specific interrupt handler *INTC* determines this by reading the status registers in *HW1* and *HW2*. It detects that *HW2* has triggered the interrupt and subsequently calls the corresponding User Interrupt Handler (*UsrInt2*).

4 Finally, *UsrInt2* calls the semaphore *Sem2*, releasing the driver code that executes in *TaskB3*. The semaphore channel uses the internal-communication services described in Section 5.5. HdS synthesis generates the interrupt code based on *UsrInt2* in the TLM.

The interrupt handler terminates after releasing semaphore *Sem2*. This finishes the interrupt sequence. As a result of the released semaphore *Sem2*, the *TaskB3*, which is pending on that semaphore, becomes ready and is subsequently scheduled. After *TaskB3* resumes execution, it reads the data from *HW2*. Finally, the process of synchronization and data transfer is finished, and the message from *B5* has arrived at *TaskB3*.

POLLING SYNCHRONIZATION

For polling based synchronization, the hardware unit exposes a memory location (a flag) to the processor bus. The hardware unit changes the value of the flag to signal synchronization. The processor periodically reads this flag in the hardware unit, detecting the value change. Then, it proceeds with the actual data transfer.

The implementation of polling-based synchronization is simpler than interrupt-based synchronization as no separate handler is needed. Instead, the polling code is part of the driver code itself. The driver accesses the slave's polling flag periodically to detect successful synchronization. It uses MAC services analogous to data transfer to access the processor bus. In addition, the polling code uses RAL services to maintain the user-selected polling period.

To ease comparison between interrupt-based and polling-based synchronization, we will repeat the same synchronization example between behavior *B5* and *TaskB3*. Now, the synchronization is implemented via polling. Figure 5.16 shows the processor portion of the polling implementation. In this figure, only the half channel *Driver* is highlighted as the synchronization is implemented inside the driver itself. Figure 5.17 outlines the sequence of events during the synchronization for sending a message from *B5* to *TaskB3*.

1 Identical to the previous example, *TaskB3* expects at t0 a message from *B5*. The driver code, executed within, *TaskB3* polls the status flag in *HW2* to

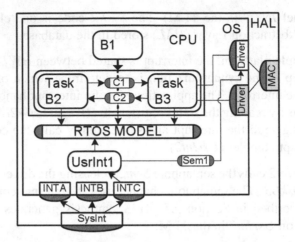

FIGURE 5.16 Polling-based synchronization

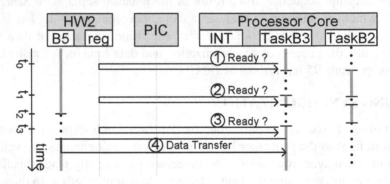

FIGURE 5.17 Events in polling-based synchronization

determine whether the message is available. Since it is not, *TaskB3* suspends for the polling period before trying again. During that time, the next lower priority ready task, in this case *TaskB2*, is scheduled onto the processor.

2 At t1, *TaskB3* re-awakes to poll the status flag again. The message is still not available and therefore *TaskB3* suspends again. Subsequently, *TaskB2* is scheduled onto the processor. Meanwhile, at t2, the message becomes available in *B5*, and the status flag changes. *TaskB3* , however, does not notice this immediately and remains suspended until the end of the polling period.

3 At t3, *TaskB3* awakes again after the expiry of the polling period. It reads the status flag in *HW2* again and detects that the message is available. Subsequently, the polling loop terminates and the task can proceed to the data transfer.

4 At t4, after successful synchronization, *TaskB3* performs the data transfer and reads the message from *B5*.

As our example has shown, polling introduces a latency between the actual availability of the message (at t2) and its detection in the transferring task (at t3). This latency is at most as long as the polling period (i.e. if the flag changes just after polling it). Using a shorter polling period reduces this latency. However, with a shorter polling period, the number of polls increases. This increases the system overhead, as the polling task has to be activated each time to read the flag in the HW unit. Hence, there is a trade-off between the maximum polling latency and the incurred system overhead.

Please note that polling without any polling delay, also called "busy waiting," should be avoided as a general solution. In the case of busy waiting, the CPU spends all its processing time for polling the status flag. Hence, it cannot schedule any lower priority tasks (such as *TaskB2* in our example). Furthermore, busy waiting leads to a pollution on the processor bus, as most of the bus capacity is exhausted by checking the polling flag. Busy waiting may effectively block not only the own processor for any other computation, but also may starve any other bus traffic on the processor bus.

In the past pages, we have described the implementation of two options for handling the synchronization necessary for external communications. Whether a developer uses interrupt-based or polling-based synchronization depends on the application and the system characteristics. Next, we will describe the MAC layer for data transfer which takes place after a successful synchronization.

5.6.4 MEDIA ACCESS CONTROL

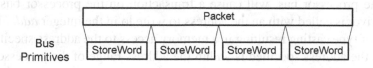

FIGURE 5.18 Transferring a packet using bus primitives

The final aspect of external communication is the actual data transfer using the MAC driver. It assumes that the previously described data formatting, packetization and synchronization are already performed. The MAC driver is the lowest layer within the external communication stack that is implemented in software. It communicates with the processor hardware which in turn implements the bus protocol to transfer data on the processor bus.

The MAC driver inside the Hardware Abstraction Layer (HAL) provides access to the bus medium. It allows transmission of packets over the processor bus. It also splits packets into bus primitives. Figure 5.18 shows an example in

```
1   void masterWrite(unsigned int addr, void *pD, unsigned int len) {
2       unsigned char *p = (unsigned char*)data;
3       while (len >= 4 ) {
4           *((unsigned int*)addr) = *((unsigned int*)pD);
5           len −= 4; pD += 4;
6       }
7       if (len >= 2 ) {/* remaning short */
8           *((unsigned short*)addr) = *((unsigned short*)pD);
9           len −= 2; pD += 2;
10      }
11      if (len >= 1) {/* the last byte */
12          *((unsigned char*)addr) = *((unsigned char*)pD);
13          len −= 1; pD += 1;
14      }
15  }
```

LISTING 5.10 MAC driver example

which a packet is split and transferred in words. The MAC driver is processor-
and bus-specific. The complexity of a MAC driver varies with implementation.
A MAC driver can be very simple if it connects to the processor bus, which is
available through memory accesses. More elaborate drivers are needed when
the targeted bus is not directly mapped to the processor memory and instead
is only accessible via a protocol transducer accessed through registers. For
example, a protocol transducer is typically used for Controller Area Network
(CAN) buses or IIC buses.

Listing 5.10 shows a very simple MAC driver designed to access the processor's data bus. It can be simple, because any memory access within the address
range the processor bus, will cause a transaction on the processor bus. The
MAC driver is called with an the address to write to in the integer *addr*. It uses
a series of type casting resulting in a memory access to the address specified by
addr. If the address specified is within the address range of the processor bus,
a bus transaction (a read or write) will be triggered on the bus.

The MAC driver is implemented in ANSI-C and triggers bus accesses through
pointer casts. The driver splits the input packet into bus transactions, as in the
while-loop in lines 3 through 6, and the subsequent conditionals in lines 7 and
11. The data transfer is accomplished by a sequence of casts. Line 4 shows
an example of writing a word to the bus. On the right hand side, the source
pointer *pD* is casted to an unsigned int pointer. Then the source pointer's value
is requested with the star operator. As a result, the right hand side contains
the value of the source data. This is assigned to the left hand side. On the left
hand side, the target address is captured in an unsigned int. It is casted to an
unsigned int pointer. Subsequently, the value of that pointer is requested with

```
1  /* processor startup code */
2  void main(void) {
3     PE_Struct_Init(&PE0);
4     BSP_init();
5     OSInit();
6
7     c_os_handshake_init(&PE0->sem1);
8     c_os_handshake_init(&PE0->sem2);
9     BSP_UserIrqRegister(INT1, Int1Handler, /*..*/);
10    BSP_UserIrqRegister(INT2, Int2Handler, /*..*/);
11
12    taskCreate(task_b2b3, NULL,
13                     B2B3_main, &this->task_b2b3);
14
15    OSStart();
16 }
```

LISTING 5.11 Startup code example

the star operator. As a result, the right hand side's value is written to the memory location identified by *addr*. Similarly, a short is written in line 8 and finally a byte in line 12. The driver for this example bus uses very similar code for reading from processor memory, with the left- and right-hand sides reversed. This is typical of simple MAC drivers.

In summary, the MAC is the lowest layer for external communication implemented in software. It provides a canonical access to the bus medium transfer data over the bus. Its implementation hides hardware-specific details, providing a generic API. The MAC driver is typically not generated, but instantiated from a database.

5.7   STARTUP CODE

Specific code is required to initialize all hardware and software components during startup of the processor. This code, sometimes called boiler plate code, is highly platform specific.The startup code must be generated within software synthesis to connect all previously generated code segments together.

Listing 5.11 shows an excerpt of the startup code. It first initializes the processor specific data structures (line 3). This sets up the structure created during code generation to represent the hierarchical composition and connectivity of the user computation (see Chapter 5.3). Next, *BSP_init()* in line 4 initializes the processor's basic support hardware and drivers (e.g. timer, PIC). This is followed by an initialization of the operating system, *OSInit()*, in line 5. This

OS-specific step sets up the operating system data structures to prepare for the instantiation of OS primitives in the user code. However, *OSInit()* does not yet start multi-tasking.

After initializing the OS, application and platform specific code sets up interrupt synchronization and create the user tasks. Lines 7 and 8 create two semaphore channels, they synchronize for external communication between an interrupt handler and the driver code as we have outlined in Section 5.6.3. Lines 9 and 10 register the corresponding interrupt handlers using *BSP_UserIrqRegister()*. The startup code generator can use the interrupt mapping information in the system TLM to connect with the appropriate external interrupt line. Line 11, creates the user task *B2B3* and with the main function *B2B3_main*. Finally, line 15, enables multi-tasking by calling *OSStart()*. At this point, the user task *B2B3* begins to execute and may dynamically create further tasks.

In short, the startup code consists of both platform-specific and application-specific code. It initializes the underlying hardware, registers interrupt handlers, and prepares multi-tasking. As a last step, it releases multi-tasking and the user tasks start executing the user defined behavior.

5.8   BINARY IMAGE GENERATION

The final aspect of SW synthesis is the generation of a complete target binary. Figure 5.19 outlines this process. Software synthesis, consisting of code generation and HdS generation, produces code for application, drivers, and interrupts. In addition, HdS generation creates build and configuration files. Using these configuration files, a cross-compiler tool chain compiles the generated code. The cross compiler is specific to the target processor and binary format. In the process, the build and configuration files select components needed for a complete target implementation from the software database and configure these components. Selected database components that are available as sources are also cross compiled into target-object files. In a final step, all object files are linked to the final target binary.

The software component database provides the essential elements for assembly of the final target binary. An effective database design is important for establishing a flexible synthesis flow, with a wide variety of configurations and many processor and hardware combinations. It is essential to identify the dependencies of each database component with respect to the selected hardware/software configuration, e.g. the selected processor, RTOS, cross compiler, and board components. Capturing all dependencies is necessary for correctly selecting a component. On the other hand, overly specializing a component may lead to code duplication within the database and yield code bloat.

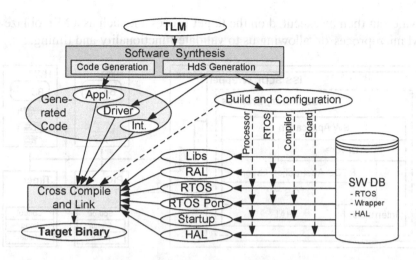

FIGURE 5.19 Binary image generation

The matrix of arrows in Figure 5.19 symbolize the dependencies when selecting a component. Usually the most specific element is the RTOS port, since it depends on the RTOS type, the processor, and the cross-compiler (all of which, for example, are necessary for the call frame layout and the stack layout required for task creation). Our software generation also produces a customized Makefile, which selects the components according to the architecture information in the TLM and then uses the cross-compiler to generate the target binary.

Automating the step of target binary generation has many advantages. It hides the complexities of the build process from the user. Using the TLM as an input for generation avoids duplication of configuration information (i.e. duplicating between the TLM and Makefiles) and allows for a tight and optimized integration with the component database. Overall, it minimizes the user effort.

Binary image generation completes the whole generation process. In the preceding sections we have described the software synthesis process, starting from task code generation, multi-tasking synthesis, communication synthesis, and the generation of startup code. Now, after the final step, the generated target binary is ready for execution.

5.9 EXECUTION

After successful target binary generation, the produced binaries are ready for download onto the target platform. The target platform may be implemented as an ASIC or using a FPGA prototyping platform. The generated embedded software can then be executed on the target processor, such as a Mircoblaze or ARM microprocessor, allowing us to validate functionality and timing.

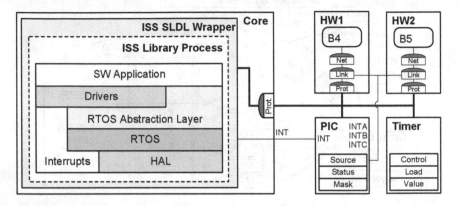

FIGURE 5.20 ISS-based Virtual platform

Alternatively, the binaries can be validated using an ISS-based virtual plat-form. Figure 5.20 depicts a processor model with an integrated ISS as part of a system TLM. Similar to the system TLM introduced earlier in this chapter, it contains a PIC and timer in an abstract form. Within the processor module, the processor core is replaced with an ISS library process. The ISS is wrapped into an SLDL wrapper. The wrapper calls the ISS cycle-by-cycle. It detects bus access requests from the ISS and translates those into calls to the abstract bus model. In the reverse direction, the SLDL wrapper listens to incoming interrupts from the system simulation, and forwards those to the ISS. The ISS interprets the generated target binary and executes the embedded software.

Both approaches of executing the target binary allow validation of the gener-ated software for its functionality and performance. If the performance analysis reveals opportunities for performance tuning, the designer can update commu-nication or computation parameters, application mapping, or even update the application specification, and then trigger the synthesis again. With the au-tomatic generation, alternative solutions can be quickly and easily generated. This allows for a rapid exploration of the embedded software design space.

5.10   SUMMARY

In this chapter, we introduced a software synthesis approach which can gen-erate C code from system models described in SLDL. The generated C code can be compiled and linked to produce a final target binary for each processor in the system.

Embedded software synthesis is an essential aspect of implementing today's complex designs. It allows us to avoid the tedious and error prone manual implementation for customized embedded systems. We have shown software generation as an integral part of an ESL flow. From the system TLM, the software synthesis automatically generates the binaries for each processor in the system. Together with the system synthesis, it completes the ESL flow for the software, offering a solution that is seamless from the abstract system model down to its implementation on embedded processors.

The presented software synthesis addresses the four aspects of creating embedded software: code generation, communication generation, multi-task generation, and binary image generation. It generates communication drivers, interrupt handlers, and adjusts for the target multi-tasking. It supports the traditional targeting toward an existing RTOS and, furthermore, offers an interrupt-based alternative for multi-tasking if an RTOS-based execution is undesirable.

Today's embedded systems are highly customized as a composition of specialized heterogeneous components. Both pre-existing IP components, as well as synthesized of application specific hardware components are combined creating application specific platforms.. Manual code development for such customized platforms is too error prone and time consuming for current market demands. Traditional software engineering approaches do not sufficiently address this problem, as they target general architectures. Therefore, customizing embedded platforms demands an automated software synthesis. Automation offers significant gains in productivity and allows the designer to focus on the essential algorithm without the burden of low-level implementation details. Automation therefore supports a shift in focus away from low-level implementation, toward a feature-oriented design.

Chapter 6

HARDWARE SYNTHESIS

HW components are synthesized as standard or custom processors or as special custom hardware units which are also called intellectual property components (IPs). As we explained in the previous chapter the synthesis process starts with specification (usually an instruction set or C code) and ends with a RTL code in an HDL that is ready for further processing with RTL tools. This synthesis process is sometimes called C-to-RTL design.

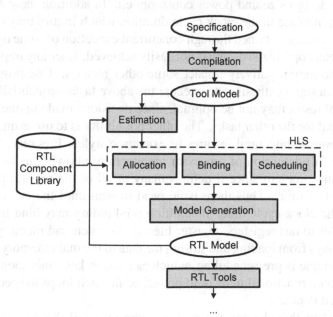

FIGURE 6.1 HW synthesis design flow

D.D. Gajski et al., *Embedded System Design: Modeling, Synthesis and Verification*,
DOI: 10.1007/978-1-4419-0504-8_6,
© Springer Science + Business Media, LLC 2009

The synthesis process starts, as shown in Figure 6.1, with a given specification, which is compiled into some intermediate tool representation, or tool model. This model can be used for estimation of different design metrics in the proposed or generated design. These metrics can also be used for some partial or complete allocation, as well as binding and/or scheduling at the start of synthesis or during design-optimization iterations.

HW component synthesis, which is usually called High-Level Synthesis (HLS), uses the tool model to estimate metrics and performs allocation, binding, and scheduling tasks. The allocation task selects necessary and sufficient components from the RTL component library and defines their connectivity or component architecture. The binding task performs variable merging and binds variables to registers, registers files, and memories, and also assigns operations to specific functional units and register-to-register transfers to available connections. The scheduling task assigns register transfers and operations to clock cycles. All of these tasks are designed to optimize design metrics such as performance, cost, size, power, testability, dependability or some other metric. These tasks must be coordinated since optimization in one of the tasks requires also support from other tasks. For example, adding an extra ALU in the datapath requires also an increase in number of ports in the register file and number of busses to supply operands to and from the newly added ALU. Furthermore, adding new resources may improve performance but it may also increase the design size and power consumption. In addition, new ALU may introduce an increase in the clock cycle duration which in turn may cancel the gain in performance obtained through concurrent execution of some operations.

A completely optimized design is not easily achieved, since any improvement in one metric may negatively impact some other metrics. One possibility of simplifying design synthesis is to execute the above tasks sequentially. In this case the final result may not be optimal, since decisions made in one task may not be optimal for the other tasks. The other possibility is to use estimation and predefine some architectural features or execution styles. Pre-allocation helps in partial or full definition of processor architecture, allowing us to avoid the timing-closure problem since it defines many or all of the register-to-register delays ahead of time. Thus there is no need to wait until the end of HLS to determine the clock cycle time. In addition pre-binding may bind frequently-used variables to fast registers, register files, or a scratch-pad memory to avoid lengthily delays from loading and storing the data to the main memory. Another helpful technique is pre-scheduling, which can assign key inner loops to high-speed pipelined functional units or to pre-schedule such loops to specific paths in a pipelined datapath.

In the rest of this chapter, we will describe in detail those tasks used for synthesis of HW components.

6.1 RTL ARCHITECTURE

RTL architecture consists of two basic components: a controller and a datapath as shown in Figure 6.2. The controller indicates the state of the architecture and provides control signals to the datapath for every clock cycle. It also receives some control inputs and outputs for coordination with other components in the platform. The datapath, on the other hand, receives the data, executes the assigned functions, and outputs the results. Each datapath also outputs status signals to the controller, which are then used to determine the next step in computation.

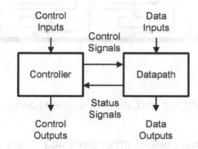

FIGURE 6.2 High-level block diagram

The more detailed RTL architecture of a controller and a datapath is shown in Figure 6.3. We can define a simple controller for the simple HW components such as memory controllers, interrupt controllers, bridges, transducers, arbiters, and other interface components with a Finite State Machine (FSM). A FSM consists of a State Register (SR) that contains the state of the FSM and two logic components: input logic and output logic. Input logic computes the next state of the FSM from the present state and the control inputs, while the output logic defines the control signals for the Datapath and control outputs from the present state and the control inputs.

A datapath contains different RTL components such as registers, register files, and memories for storage of data, as well as different functional units for computation, such as ALUs, and the MULs. Each storage and functional unit can take one or more clock cycles and can be pipelined in one or more stages. These units can be connected with busses or with point-to-point connections through selectors. Of course, some of the units can be chained so that data from one unit to the other unit goes directly or through a register. Each unit may have input and output registers for storing temporary data or for data forwarding. Some or all register-to-register paths can be pipelined so that several different operations can be executed concurrently in different pipeline stages, although each operation takes approximately the same amount of time to execute.

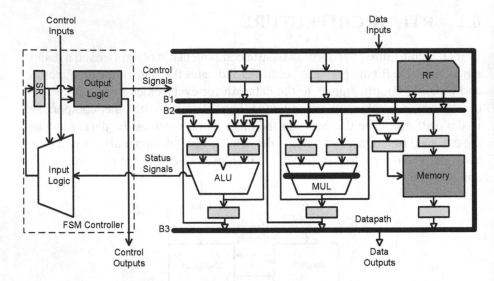

FIGURE 6.3   RTL diagram with FSM controller

For larger standard and custom processors and larger special function processors, the simple FSM controller is usually replaced with a programmable controller, as shown in Figure 6.4. In this case, the State Register becomes the Program Counter (PC); the output logic becomes the Control Memory (CMem) for storing control words, or the Program Memory (PMem) for storing instructions; and the input logic becomes the Address Generator (AG) for generating the address of the next control word or the next instruction. The AG may compute the next address from information supplied by different sources, such as a datapath address, datapath status signals, an offset from the CW or an instruction and data supplied by the rest of the platform through control inputs.

This type of programmable controller can be also pipelined by itself or in conjunction with the datapath. Figure 6.4 shows such a pipelined controller together with a pipelined datapath. In order to pipeline the controller we may insert an Instruction Register (IR) or a Control Word Register (CWR) to store temporarily instructions or control words. We can also insert a Status Register (SR) to store status signals from the datapath. After insertion of these registers, every instruction or control word as shown in Figure 6.4 needs three cycles to be generated before applied to the datapath. In the first cycle, AG generates the new address for PC from information in registers such as SR or IR or CWR or some other registers in the datapath. In the second cycle a control word from CMem or instruction from PMem is loaded into IR or CWR. Finally, in the third cycle control word or decoded instruction is applied to the datapath. Furthermore, the number of cycles may increase if CMem or PMem read takes more than one clock cycle...

In addition to controller pipelining, datapath can be pipelined at the same time. For example, the datapath in Figure 6.4 has input and output registers preceding and following every functional unit. Therefore, it takes one clock cycle to read the data from the RF and transfer it to the ALU input register, one clock cycle to compute the operation in the ALU and store date in the ALU output register, and finally one clock cycle to write data back into the RF. Of course, some functional units may be pipelined and take more than one clock cycle to complete their operation. For example multiplier MUL in Figure 6.4 has two pipelined stages and will take two clock cycles to complete multiplication of two operands in its input registers before the result is stored in the MUL output register. Therefore, data computation from RF to RF in the datapath of Figure 6.4 may take three (arithmetic and logic operations) or four (multiplication) clock cycles. With pipelined controller and datapath as shown in FIGURE 6.4, new address generation takes at least four clock cycles: one from the PC to the IR or CWR, the second one to the ALU input registers, the third to the SR or ALU output register, and finally, the fourth one back to the PC. It may, however, take more than four clock cycles if a new address is delivered from the local memory which takes longer than one clock cycle to read the address.

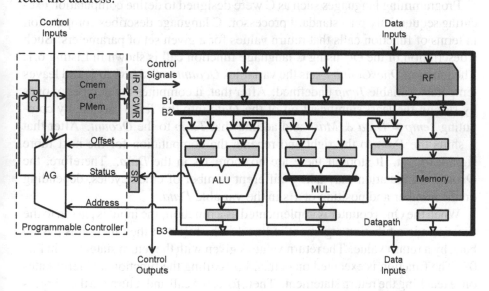

FIGURE 6.4 RTL diagram with programmable controller

6.2   INPUT MODELS

Input models for HW component synthesis come in many different forms: from C-code to RTL to netlists. The C-code input does not contain any design decisions, while logic netlist, for example, has all the design decisions already implemented. Many other forms, such as CDFG or FSMD, include some but not all of the design decisions. We will look at several different input models using the example of the Ones Counter (OC). The OC takes in *Data* as a string of 0's and 1's, computes the number of 1's in the *Data*, and then outputs this number of 1's after certain amount of time. The OC waits for the *Data* until the *Start* signal becomes 1, and then copies the *Data* and starts counting. It stops when all the ones are counted, then outputs the number of ones, sets the *Done* signal to 1, and goes to the waiting state, in which it waits for the *Start* signal to go to 1 again. The OC example can be used to describe different input models for HW component synthesis.

6.2.1   C-CODE SPECIFICATION

Programming languages such as C were designed to define computation executing sequentially on a standard processor. C language describes computation in terms of function calls that return values for a given set of parameters. Such a description of the OC using C language function call is shown in Listing 6.1. The function *OnesCounter* sets the variables *Ocount* to 0, *Mask* to 1, and leaves temporary variable *Temp* undefined. After that, it computes the number of ones by storing the least significant bit of the *Data* into the *Temp* variable by computing *Temp = Data & Mask* and adding that *Temp* to the *Ocount*. After that it shifts the *Data* to the right and repeats the computation for the next more significant bit. It stops if there are no more 1's in the *Data*. Therefore, the *Ocount* computation may take a different number of clock cycles, depending on the number and position of 1's in the variable *Data*.

When the Ones counter is implemented as a function, the input is passed to the function via a function argument in line 1. The result of the function is passed back by a return value. The return value is given with the return statement in line 09. The function is executed on demand by calling the function and terminates once reaching the return statement. Therefore, the call and return are the triggers for starting and ending the computation. In this way, the function-based C-code definition differs from a typical hardware component, which is always running and waits for control signals to start computing while it communicates through data signals with other components.

In order to describe operation of a HW component on RT level more accurately, we need to introduce new variables for controller and datapath inputs and outputs. Furthermore, we need to rewrite the C code in Listing 6.1 to execute

```
1  while(1) {
2    while(Start == 0);
3    Done = 0;
4    Data = Input;
5    Ocount = 0;
6    Mask = 1;
7    while(Data > 0) {
8      Temp = Data & Mask;
9      Ocount = Ocount + Temp;
10     Data >>= 1;
11   }
12   Output = Ocount;
13   Done = 1;
14 }
```

```
1  int OnesCounter(int Data){
2    int Ocount = 0;
3    int Temp, Mask = 1;
4    while(Data > 0) {
5      Temp = Data & Mask;
6      Ocount = Data + Temp;
7      Data >>= 1;
8    }
9    return Ocount;
10 }
```

LISTING 6.1 Function-based C code   LISTING 6.2 RTL-based C code

indefinitely. This is shown in Listing 6.2, which executes forever in the loop shown in line 1, with loop in line 2 waiting for *Start* to become 1. The C code in Listing 6.2 has several new control and data variables added. Control signal *Done* is reset to 0 in line 3 and set to 1 in line13 indicating the beginning and ending of the component operation. The variables *Input* and *Output* represent the data ports to the component while variables *Data*, *Ocount*, *Mask* and *Temp* are temporary variables for computation inside the component.

In general, designers need to avoid function calls to ensure better understanding and easier synthesis of hardware components. As a result of the re-coding shown in Listing 6.2, we now have flat C-code that is suitable for high-level synthesis. The code uses *Input* and *Output* variables for communication with the external environment and does not contain any function calls. However, we still can not make distinction between control and data variables in the code in Listing 6.2.

6.2.2 CONTROL-DATA FLOW GRAPH SPECIFICATION

A Control-Data Flow Graph (CDFG) [151] represents a C-code with if, loop, and Basic Block (BB) constructs, where a BB represents a sequence of programming statements without ifs and loops. A CDFG is useful because it shows control and data dependences. It allows us to easily follow control dependences between if, loop, and BB constructs in the C-code since each of them is represented by a graphical symbol with the control arrows between them. It also allows us to expose concurrency of operations in the C code by representing each BB as a Data-Flow Graph (DFG) in which every operation is represented by a node in the graph while arrows between the nodes represent

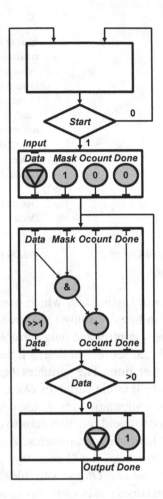

FIGURE 6.5 CDFG for Ones counter

input and output variables. Note that the order in which the statements are written in the C code imposes unnecessary sequentially on the execution of these statements.

A CDFG for the OC is shown in Figure 6.5. It consists of four BBs and two if statements. The first BB is empty, while the OC is waiting for the Start signal to become 1. The second BB does the *Data*, *Mask*, *Ocount*, and *Done* variable initialization. The third BB is executed in the loop that a count 1's until *Data* is equal to 0. It extracts the least significant bit by masking the rest of the bits away and ads it to *Ocount* after which it shifts the *Data*. The last BB outputs the *Data* and *Done* values. As in the C-code in Listing 6.2, the CDFG model does not distinguish between controller and datapath outputs. Without this distinction we may end up with an inefficient implementation in which controller outputs are implemented as datapath outputs and vice versa.

However, CDFG can represent a waiting state in which OC waits for *Start* signal to become 1, similarly to the C code in Listing 6.2 in line 2.

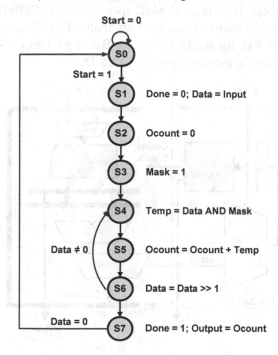

Start = 0

S0

Start = 1

S1   Done = 0; Data = Input

S2   Ocount = 0

S3   Mask = 1

S4   Temp = Data AND Mask

Data ≠ 0   S5   Ocount = Ocount + Temp

S6   Data = Data >> 1

Data = 0   S7   Done = 1; Output = Ocount

FIGURE 6.6 FSMD specification

6.2.3 FINITE STATE MACHINE WITH DATA SPECIFICATION

A Finite State Machine with Data (FSMD) [63, 184] specification is an upgraded version of the well-known Finite State Machine representation providing the same information as the CDFG specification. A FSMD specification, however, can be more specific than the corresponding CDFG specification if we assume that each state is executed in one clock cycle. This way a FSMD specification can provide an accurate estimation of the design performance. The FSMD specification for the OC is shown in Figure 6.6. It has eight states and requires eight clock cycles to execute. The second BB from the CDFG specification that contains four initialization operations is executed in states S2, S3, and S4 and therefore requires three clock cycles. Since *Done* is a control signal, it can be executed concurrently with operations on any of the other three variables. Variables *Data*, *Ocount* and *Mask*, which are presumably in the same RF, need three states or three clock cycles to execute since their initialization goes through the same ALU in the implementation. This is also true of the computation of the *Temp*, *Ocount*, and *Data* values in states S4, S5, and S6 in the 1's counting loop. Therefore, a FSMD specification can offer more accurate timing information in terms of clock cycles than a CDFG specification. The main reason for this is that the DFG in each BB in a CDFG is scheduled into states or clock cycles in the corresponding FSMD.

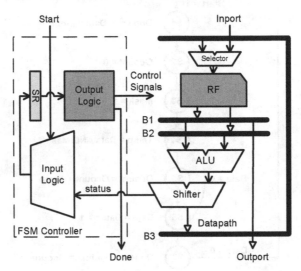

FIGURE 6.7   RTL Specification

6.2.4   RTL SPECIFICATION

Register-Transfer-Level (RTL) specification provides a Datapath netlist with all RTL components and their connections. It also includes two tables for synthesis of input logic and output logic components in the FSM Controller. We can automatically obtain Boolean equations for synthesis of input logic from the input logic table after we define some state encoding for the states in the table. Similarly, we can automatically obtain Boolean equations for the output logic from the output logic table which defines variable bindings to storage elements and operations to functional units.

For example, RTL specification for the OC is given in Figure 6.7, and in Table 6.1 and Table 6.2. In the ones counter architecture, shown in Figure 6.7, variables *Data*, *Mask*, *Ocount*, and *Temp* are in the registers *RF[0]*, *RF[1]*, *RF[2]*, and *RF[3]* of the two port register file *RF*. They communicate through buses *B1* and *B2* with two chained functional units, *ALU* and *Shifter*. *Shifter* also provides the status signal *Data = 0* to the input logic in the FSM Controller. The input logic table in Table 6.1 supplies logic equations for the next state and the control output signal *Done*. We can see that *Done* = *1* when the SR is in state S7. Similarly, we can derive Boolean equations for the output logic from the output logic table shown in Table 6.2 once control encoding for every storage and functional unit is taken into account.

TABLE 6.1 Input logic table

Present State	Inputs: Start	Data = 0	Next State	Output: Done
S0	0	X	S0	X
S0	1	X	S1	X
S1	X	X	S2	0
S2	X	X	S3	0
S3	X	X	S4	0
S4	X	X	S5	0
S5	X	X	S6	0
S6	X	0	S4	0
S6	X	1	S7	0
S7	X	X	S0	1

TABLE 6.2 Output logic table

RF[0] = Data
RF[1] = Mask
RF[2] = Ocount
RF[3] = Temp

State	RF Read Port A	RF Read Port B	ALU	Shifter	RF selector	RF Write	Outport
S0	X	X	X	X	X	X	Z
S1	X	X	X	X	Inport	RF[0]	Z
S2	RF[2]	RF[2]	subtract	pass	B3	RF[2]	Z
S3	RF[2]	X	increment	pass	B3	RF[1]	Z
S4	RF[0]	RF[1]	AND	pass	B3	RF[3]	Z
S5	RF[2]	RF[3]	add	pass	B3	RF[2]	Z
S6	RF[0]	X	pass	shift right	B3	RF[0]	Z
S7	RF[2]	X	X	X	X	disable	enable

6.2.5   HDL SPECIFICATION

RTL specification in VHDL/Verilog provides a FSMD description from which we can derive a datapath netlist with all RTL components and their connections, as well as a FSM description for logic synthesis of the input logic and output logic.

```
1  // ...
2  always@(posedge clk)
3  begin : output_logic
4    case (state)
5      // ...
6    S4: begin
7      B1 = RF[0];
8      B2 = RF[1];
9      B3 = alu(B1, B2,l_and);
10     RF[3] = B3;
11     next_state = S5;
12   end
13     // ...
14   S7: begin
15     bus_32_0 = RF[2];
16     Outport <= B3;
17     Done <= 1;
18     next_state = S0;
19   end
20   endcase
21 end
22 endmodule
```

LISTING 6.3   RTL description in HDL (excerpt)

For example, Listing 6.3 shows a Verilog description of the OC presented in
Figure 6.7. It states that the case statement on line 3 always gets executed on
the positive edge of the clock signal. The case statement represents Table 6.1
and Table 6.2. For example, in state *S4*, variables *Data* and *Mask* in *RF[0]* and
RF[1] are supplied through buses *B1* and *B2* to ALU. ALU performs an AND
operation and outputs the result *Temp* to bus *B3*. *Temp* from bus *B3* is written
back into register file *RF* at location *RF[3]*. After that the OC controller goes
to state *S5*. Similarly, *Ocount* from *RF[2]* is sent to *Outport* is state *S7*.

We can see that Verilog distinguishes between variables and signals. Verilog
uses = for variable assignment and <= for signal assignment. In many ways,
the Verilog description in Listing 6.3 is similar to the FSMD description in
Figure 6.6. It has the same number of states and the same transitions from state
to state. However, Verilog description is more detailed then FSMD description
since its variables reflect components and buses in the datapath netlist.

6.3   ESTIMATION AND OPTIMIZATION

HLS starts with the selection of some initial RTL components such as storage components, functional units, and buses. The datapath then executes variable assignments in every clock cycle, a process by which selected variables will be assigned new values through arithmetic, logic, and shift operations that are performed by functional units. To execute each variable assignment statement, then, the datapath must take data from a storage component that stores the variables in the right-hand side of the assignment, pass this data to the functional units that compute the new value, and then pass it back to the storage component, which stores the variables on the left-hand side of the equation. Given this process, it follows that we can approach datapath creation by selecting the storage components, the functional units, and the buses that connect these components.

By focusing on the storage components, for example, we note that the variables in the datapath must be stored in registers, register files, and memories. However, since not all variables are alive at the same time, it is possible for certain variables to share the same register or the same location in a register file or a memory. In other words, we can merge the datapath variables in a way that reduces the number of storage locations in the datapath. Furthermore, even if certain variables are alive at the same time, they may not be accessed at the same time, which means that we could combine them into a register file or a scratch-pad memory so that they can share the same register file or memory ports. In this manner, by combining storage locations we minimize the number of storage ports in the datapath and thus reduce the number of connections needed.

We can reduce the number of functional units in the datapath in a similar way. As mentioned above, in each state, selected variables are to be assigned new values through various arithmetic, logic, or shift operations, each of which can be performed by a separate functional unit. However, since most of these operations are executed in different clock cycles, they could share the same functional unit. In other words, we can reduce the number of units in the datapath by combining different operations into groups, allowing each group of operations to be executed in a single functional unit.

The third basic technique for optimization focuses on the datapath connectivity. As mentioned above, the execution of an assignment statement requires that data pass from one storage component to the functional unit that computes the new value and then back to another storage component. The data, in other words, is passed through connections between storage and functional units. However, since different connections will be used in different states, we can group connections into buses, which enable us to reduce the number of wires in the datapath.

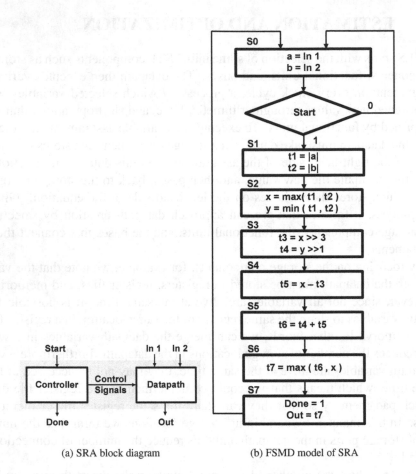

S0
a = In 1
b = In 2

Start   0

S1   1
t1 = |a|
t2 = |b|

S2
x = max(t1 , t2)
y = min (t1 , t2)

S3
t3 = x >> 3
t4 = y >>1

S4
t5 = x – t3

S5
t6 = t4 + t5

S6
t7 = max (t6 , x)

S7
Done = 1
Out = t7

Start   In 1   In 2

Controller —Control Signals→ Datapath

Done   Out

(a) SRA block diagram   (b) FSMD model of SRA

FIGURE 6.8 Square-root algorithm (SRA)

The components can be selected by assuming that one operation is executed every clock cycle or that all possible operations not dependent on each other are executed in the same clock cycle. In the first case, we have a cost-driven design, while in the latter, we have a performance-driven design.

These three basic allocation techniques can be demonstrated on a small example dedicated to computing of the Square-Root Approximation (SRA) of two signed integers, a and b, by the following formula:

$\sqrt{a^2 + b^2} \approx max((0.875x + 0.5y), x)$

where $x = max(|a|, |b|)$ and $y = min(|a|, |b|)$.

According to the FSMD model in Figure 6.8(a) this component has two input ports, In1, and In2, which are used to read integers a and b, and one output port *Out*. As you can see in the FSMD model in Figure 6.8(b), the component reads the input ports and starts the computation whenever the input control signal *Start* becomes equal to 1. In state *S1*, it computes the absolute values of variables *a* and *b*, and in *S2*, it assigns the maximum of these two values to *x* and the minimum to *y*. In state *S3* it shifts *x* three positions to the right to obtain *0.125x* and shifts *y* one position to the right to obtain *0.5y*. The SRA component calculates *0.875x* by subtracting *0.125x* from *x* in state *S4*. In state *S5* it adds *0.875x* and *0.5y*, while in state *S7* it computes the maximum of *x* and the expression *0.875x + 0.5y*. In state *S7*, the SRA component produces the result and makes it available through the *Out* port for one clock cycle. At the same time, it sets the control signal *Done* to 1, in order to signal to the environment that the data that has appeared at the *Out* port is a valid result.

To determine resource requirements from this FSMD model, we would need to generate the variable and operation usage tables shown in Table 6.3 and Table 6.4. In the variable-usage table, each row represents one variable found in the FSMD model and each column represents one state. For each variable, then, we would enter an x in the columns that correspond to the states in which the variable is alive. A variable is considered alive in the first state that follows the rising edge of the clock signal which assigns its new value and also in all states inclusively between the first and final states in which this new value is used for the last time. In Table 6.3, for example, variables *a* and *b* are assigned their values at the rising edge of the clock signal indicating the beginning of states, but they are not used in any other states.

TABLE 6.3 Variable usage

	S1	S2	S3	S4	S5	S6	S7
a	X						
b	X						
t1		X					
t2		X					
x			X	X	X	X	
y			X				
t3				X			
t4				X	X		
t5					X		
t6						X	
t7							X
No. of live variables	2	2	2	3	3	2	1

Therefore, variables *a* and *b* are alive only in state S1. By contrast, variable *x* is assigned its new value at the beginning of state *S3*, but the value of *x* is also used in states *S4* and *S6*, indicating that the variable *x* is alive in states *S3*, *S4*, *S5*, and *S6*. On the basis of this table, then, we can see which variables are alive in which states.

TABLE 6.4 Operation usage

	S1	S2	S3	S4	S5	S6	S7	Max. no. of units
abs	2							2
min		1						1
max		1				1		1
>>			2					2
-				1				1
+					1			1
No. of operations	2	1	2	1	1	1		

More importantly, however, Table 6.3 also shows the maximum number of variables alive in a single state. That is, it shows us that in states *S4* and *S5* there are three live variables. We would therefore conclude that we will need at least three registers in the datapath of this SRA design. Because of this, we may be able to combine variables from Table 6.3 into three groups so that each group that is to be stored in one of the registers contains only variables that are not alive at the same time. On the basis of this example, then, you can see that one of the major tasks in RTL synthesis consists of merging or grouping variables and assigning the groups to registers or memory locations in a way that will minimize the number of storage components or some other design metric, such as performance, power, or testability. Since each group of variables shares a register or memory location, this task is also frequently called register/memory sharing.

In a similar fashion, we might determine the minimum number of units needed to execute all the operations in the design. For this purpose we would use Table 6.4, in which the rows represent the different operator types found in the FSMD model and the columns represent the states, as before. From this table we can conclude that we need two units that can compute absolute value (indicated by | | in the FSMD model) and shift data (indicated by ») and one unit that can perform *max*, *min*, +, and - operations. Given these requirements, the straightforward approach to designing the datapath for the SRA is to allocate two units for computation of absolute value, two shifters, and one unit each for the computation of maximums and minimums, one adder, and one subtractor. The problem with this straightforward implementation, however, is that we do not necessarily need one functional unit per operation. Since no state uses all of these operations simultaneously, the implementation of one unit per operation will have functional units idling most of the time. In fact, we do not need more than two operations in any one state, so it is more efficient to construct functional units that can perform more than one operation, as this allows a substantial hardware saving.

For example, in the SRA description, addition and subtraction are never performed at the same time, which means that we can merge these operations into one functional unit called an adder/subtractor. In this case we gain one adder and a complementer at the expense of an additional EX-OR logic. On the other hand, merging the 1-bit shifter and the 3-bit shifter does not save hardware but requires an additional selector. On the basis of these examples, you can see how we perform the second major task in RTL synthesis, which consists of merging or grouping operators and designing a functional unit for each group, thereby minimizing a given design metric such as area, the number of gates or transistors, or the number of functional units in the datapath. This task is also called functional-unit sharing.

TABLE 6.5 SRA connectivity

	a	b	t1	t2	x	y	t3	t4	t5	t6	t7
abs1	I		O								
abs2		I		O							
min			I	I	O						
max			I	I	I	O			I	O	
>>3						I		O			
>>1							I		O		
-							I	I			O
+									I	I	O

If our primary goal is to minimize wiring, we should also consider merging connections into buses, since each single connection between any two units would be used in very few states and would mostly remain idle. As an example, let us consider the connections in an SRA datapath that uses one register per variable and single-operation functional units. The connections for such a datapath are given in the connectivity table shown in Table 6.5, in which each row corresponds to one functional unit and each column represents one register. To complete the table, we enter the letter I for every connection between a register and the input of a functional unit, and for every connection between the output of the functional unit and a register, we enter the letter O. As you can see in Table 6.5, such an SRA requires 14 input connections and 9 output connections, for a total of 23 connections. Of these 23 connections, however, very few are needed in any one state. From the FSMD model, in fact, we know that the maximum number of connections is used in state *S2* is four input connections, which link the registers storing the variables *t1* and *t2* to the *min* and *max* units as well as two output connections, linking *min* and *max* units to the registers that store the variables *x* and *y*. In other words, the maximum number of connections needed concurrently is six.

From this example, you can see that the third major task in RTL synthesis consists of merging or grouping connections and assigning one bus to each group so as to minimize the connection cost. Note that this connection cost includes the cost of bus drivers, which are required for every connection of a unit to a bus, and the cost of input selectors, which are required whenever two or more buses are connected to the same input of a storage or functional unit. This task is also frequently called bus sharing.

6.4 REGISTER SHARING

As we mentioned earlier, one of the major tasks in datapath optimization involves grouping variables so that they share a common register or memory location. The advantage of such grouping lies in the fact that it reduces the number and size of storage components, which in turn reduces the silicon area and therefore the cost of fabrication. Since a register can be shared only by those variables with non-overlapping lifetimes, this technique requires us to determine the lifetimes of each variable.

The lifetime of a variable is defined as the set of states in which that variable is alive, which includes the state following the state in which it is assigned a new value (write state), every state in which it is used on the right-hand side of an assignment statement (read state), and all the states on each path between the write state and a read state. Note also that each variable may have multiple assignments and that each assigned value may be used several times. Once we determine the lifetime of each variable, we can group variables that have non-overlapping lifetimes and assign each group to a single register.

When we group variables, one of the common goals is to try to have as few registers as possible, which means that we would try to partition variables into the smallest number of groups while ensuring that every variable belongs to one of these groups. This goal can be accomplished by a simple algorithm such as left-edge algorithm, which tries to pack as many variables as possible into each register. Left-edge algorithm is simple and fast, but it does not take into account the overall datapath structure.

As we demonstrated in the previous section for the SRA example, we cannot reduce the number of registers in the datapath to fewer than three. However, since there are many possible datapath designs with three registers, we would like to select one that minimizes a second design metric, such as connectivity cost. For example, the cost of connecting I/O ports, registers, and functional units can be measured in the number of selector inputs, assuming that the cost per selector input is constant.

To develop an algorithm that will minimize the number of registers as well as connectivity cost, we give priority to the combining of certain variables.

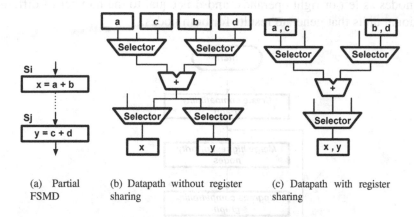

(a) Partial FSMD
(b) Datapath without register sharing
(c) Datapath with register sharing

FIGURE 6.9 Gain in register sharing

Priority is given to two variables that are used as the left or right operands for the same operator type and to variables whose value is generated by the same operator type, since merging such variables can potentially save one selector input. This concept is demonstrated in Figure 6.9(a) for two additions ($x = a + b$ and $y = c + d$) performed in different states on different operands and assigned to different variables. If we assume that both additions may be executed in the same functional unit, merging operands and results may result in the saving of selector inputs. For example, if we assign each variable to a separate register, we may obtain the design shown in Figure 6.9(b), which requires 10 selector inputs. However, if we merge variable a with c, b with d, and x with y, then assign each pair to the same register, we reduce number of selector inputs by three, as shown in Figure 6.9(c).

In general, for any n variables that are used as a source or a destination to the same operator or functional unit, there is a potential saving of n -1 selector inputs when these n variables share the same register. To consider this potential saving during variable merging, we present a partitioning algorithm, which partitions a variable compatibility graph. Such a compatibility graph consists of nodes and edges in which each node represents a variable and each edge between two nodes represents compatibility or incompatibility in merging variables represented by these two nodes. There are two types of edges in the graph: an incompatibility edge (represented by a dashed line) between two nodes indicates variables with overlapping lifetimes, while a priority edge between two nodes indicates variables with non-overlapping lifetimes that serve as the source or destination to the same functional units. Each priority edge has a priority weight indicating the number of selector inputs that can be saved. The priority weight has the form s/d, where s is equal to the number of different functional units that use both nodes as left or right operands, and d is equal to the number of different functional units that generate results for both nodes.

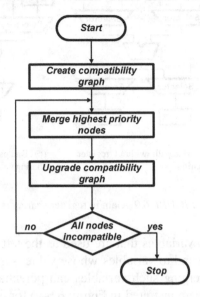

FIGURE 6.10 General partitioning algorithm

In what follows, we describe a graph-partitioning algorithm that merges compatible nodes into supernodes until all nodes in the graph are incompatible. More precisely, the algorithm always merges two nodes that are connected with a priority edge with largest weight and creates from them a supernode. Next, it deletes all the edges within the supernode and creates new edges between the supernode and other nodes. For example, it creates an incompatibility edge for any node that is incompatible with at least one node in the supernode. Conversely, it creates a priority edge for any node that is used as a common source or destination with at least one node in the supernode and that is compatible with all the nodes in the supernode. The weight of the new priority edge is computed as before. This procedure is summarized in Figure 6.10.

If we apply this algorithm to the SRA example, we obtain a grouping of variables that is slightly different from the grouping we would have obtained with some other algorithms. First, we need to create a compatibility graph, as shown in Figure 6.11(a). Note that all the variables that have overlapping lifetimes have been connected with a dashed line, indicating that they cannot be merged. To create priority edges that indicate compatibility, we assume a simple library that includes units for computing absolute value, minimum, maximum, shift, sum, and difference, in addition to functional units capable of performing a combination of operations, such as an adder/subtractor, a min/max unit, or a two-way shifter. Assuming that the units defined in this library will

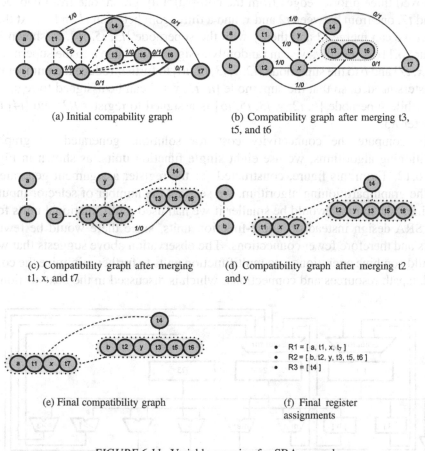

(a) Initial compability graph

(b) Compatibility graph after merging t3, t5, and t6

(c) Compatibility graph after merging t1, x, and t7

(d) Compatibility graph after merging t2 and y

(e) Final compatibility graph

(f) Final register assignments

- R1 = [a, t1, x, t7]
- R2 = [b, t2, y, t3, t5, t6]
- R3 = [t4]

FIGURE 6.11 Variable merging for SRA example

be used in the final design, we find that there are priority edges between the variables *t1*, *t2*, and *x*, and *t1*, *t2*, and *t6*. Since they are all inputs of the same max unit; there are priority edges between the variables *x*, *y*, and *t7* because they are possible destinations of a min/max unit. There are also priority edges between t3 and t5, and t5 and t6 because they all are possible inputs and outputs of an adder/subtractor.

After creating this compatibility graph, we can start merging variables and creating supernodes. In this case, all the priority edges have the same weight, so we first select those nodes whose merging will not remove any priority edges from the compatibility graph. In other words, we merge the variables *t3*, *t5*, and *t6* for a possible gain of two selector inputs, thereby creating the supernode [*t3*, *t5*, *t6*], as shown in Figure 6.11(b). Next, we select the node that has a maximum number of priority edges, namely *x*, and merge it with *t7* and then *t1* as shown in Figure 6.11(c). Note that by merging *x*, *t7*, and *t1*, we have removed three priority edges from the compatibility graph, one from between *y* and *t7*, one from between *t2* and *x*, and a third from between *t1* and *t6*. At this point we can merge *t2* and then *y* with the supernode [*t3, t5, t6*], as shown in Figure 6.11(d). Finally, we can randomly assign the variable *a* to the supernode [*t1, x, t7*] and *b* to the supernode [*t2, y, t3, t5, t6*] to further reduce the number of registers needed, so that the supernode [*a, t1, x, t7*] can be assigned to register *R1*, while supernode [*b, t2, y, t3, t5, t6*] is assigned to register *R2*, and [*t4*] to register *R3*.

To compute the connectivity cost for solutions generated by graph-partitioning algorithms, we use eight single function units, as shown in Figure 6.12. From this figure, constructed for the register assignment generated by the graph-partitioning algorithm, we see that the number of selector inputs is nine. The number would be smaller if we had used multifunctional units for the SRA design instead of single-function units, since there would be fewer units and therefore fewer connections. The observation above suggests that we should combine operations into multifunction units to further minimize the cost of datapath resources and connections, which is discussed in the next section.

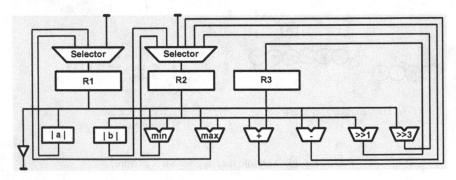

FIGURE 6.12 SRA datapath with register sharing

6.5   FUNCTIONAL UNIT SHARING

The main goal behind functional-unit sharing, or operator merging is to minimize the number of functional units in a datapath. Like register sharing, functional-unit sharing is possible because within any given state, a datapath will not perform every operation. Therefore, similar operators can be grouped into a single multifunction unit that will be used more frequently, thus increasing the unit utilization. In some cases, of course, grouping operations in this manner may not reduce the cost of the datapath; since dissimilar operators often require structurally different designs, grouping them can sometimes result in no gain or even in a higher cost.

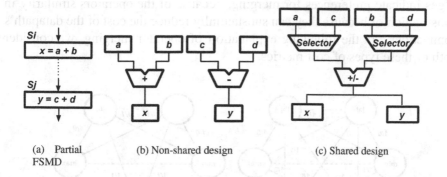

(a) Partial FSMD   (b) Non-shared design   (c) Shared design

FIGURE 6.13 Gain in functional unit sharing

In many cases, however, operator merging can yield cost reductions that are not negligible, as demonstrated in Figure 6.5. In this example, we have assumed that the datapath will perform two different operations, addition and subtraction, on different operands in different states, as indicated in Figure 6.13(a). If we implemented a partial FSMD in Figure 6.13(a) using single-function units, we would get the design shown in Figure 6.13(b), in which the datapath requires both an adder and a subtractor. We could, however, obtain the same functionality by using only one adder/subtractor and two selectors, as shown in Figure 6.13(c). Obviously, the second design would be preferable when the cost of an adder/subtractor and two selectors is less than the cost of a separate adder and subtractor. It is in cases like this that functional-unit sharing would be advantageous. Thus we would like to develop an algorithm that will combine operators into functional units in such a way that the total cost of all multifunction units and necessary selectors is minimal. For this purpose, we can use the graph-partitioning algorithm presented in the previous section. We demonstrate operator merging with this algorithm on the SRA example. For this example, we assume the availability of a complex component library that includes several multifunction units that can each compute three of more of the following operations: absolute value, minimum, maximum, sum, and difference.

To merge the operators called for in the FSMD model, we must first construct a compatibility graph that indicates which operators can be combined. Each node in the compatibility graph represents one operator type from the FSMD model, although each graph may have several nodes for each operator type. As a rule, the number of nodes will be equal to the maximum number of occurrences of a particular operator type in any single state. To indicate the compatibility of the various operators, we need to connect the nodes in the graph with priority edges or incompatibility edges. As you would expect, an incompatibility edge indicates that its two operators cannot be merged under any circumstances, since they are to be used concurrently in the same state. By contrast, the priority edges indicate preferences for merging, because of the operators'similarity in construction or because they can substantially reduce the cost of the datapath's connections. In the following explanation of operator merging, we consider both of these types of cost metrics.

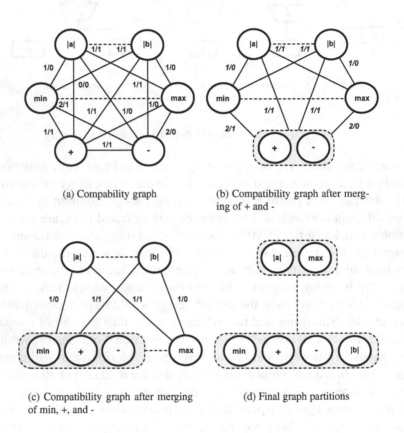

(a) Compability graph

(b) Compatibility graph after merging of + and -

(c) Compatibility graph after merging of min, +, and -

(d) Final graph partitions

FIGURE 6.14   Functional unit merging for SRA

In creating the compatibility graph shown in Figure 6.14(a), we excluded shift operators, because their cost is zero and grouping them with other operators would only increase the cost of the SRA datapath. As you can see, we included two absolute-value operators in the graph, since the absolute values of *a* and *b* are to be computed simultaneously in state *S1*. Including the remaining operators, we find that the graph has six nodes and two incompatibility edges: One connects the two absolute-value operators since they are to be used in the same state, while the second edge connects the maximum and minimum operators since they are incompatible for the same reason.

If we had used single-function units, then implementing the SRA datapath would require two absolute-value units and one unit each for maximum, minimum, addition, and subtraction. Intuitively, the total cost of all these units would be too high. Therefore, we need to use complex functional units that can do several different arithmetic operations. By merging operators in the compatibility graph, we can reduce the datapath cost in several different ways. Looking into compatibility graph we see two alternatives. One is to merge |a| and *min* operators in one group, and |b|, *max*, +, and - into another multi-function unit. The other alternative is to merge |a|, *min*, and + in one, and |b|, *max*, and - operators into another functional unit. Note that either one of these alternatives would cost much less than the original implementation using single-function units. In general, the designs generated by operator merging have much lower functional-unit costs, which means that as a whole, these designs are more cost-efficient.

However, it is also possible to reduce the datapath cost further by minimizing connectivity cost while merging operators. To achieve this, we must use the priority edges in the compatibility graph, as we did in the case of variable merging. Again, the weight of these priority edges is based on the number of common sources and common destinations.

Returning to the compatibility graph shown in Figure 6.14(a), we can now add priority edges and redesign the SRA datapath so as to reduce the number of selector inputs while merging operators. In Figure 6.14(a), for example, we have labeled each priority edge with a weight s/d, in which s indicates the number of common sources and d indicates the number of common destinations. As you can see, the edge between the + and - operators is labeled 1/1, since two source variables (right operands), $t3$ and $t5$, and two destination variables, $t5$ and $t6$, share register $R2$. Similarly, the edge between the *min* and - operators is labeled 2/1, since *min* and - have two common sources and one common destination, that is, the left operands, $t1$ and x, share register $R1$, the right operands, $t2$ and $t3$, share register $R2$, and the results, y and $t5$, share register $R2$.

At this point we can use the graph-partitioning algorithm presented in Figure 6.10 to group these operators into the appropriate functional units. According to this algorithm, we first try to group those operators that have a similar design structure, such as addition and subtraction, *min* and *max*, and left shift and right shift. In general, grouping similar operators in this manner will produce the largest cost reduction. In the case of the SRA algorithm, for example, we could group the + and - operators into a single supernode and then redraw the compatibility graph as shown in Figure 6.14(b). Next, we add the *min* operator to this supernode, since it has the largest number of common sources (two) and destinations (one) of all the nodes in the graph. This next version of the compatibility graph is shown in Figure 6.14(c). Finally, we add to this supernode the absolute-value operator for variable b, for the same reason, and then merge the max operator with the absolute-value operator for variable a. At this stage we have arrived at the graph partition shown in Figure 6.14(d), which cannot be reduced further.

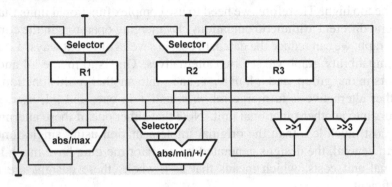

FIGURE 6.15   SRA design after register and unit merging

As you can see from this partitioned graph, we should be able to construct a datapath for the SRA algorithm by using three registers and four functional units. The final assignment of the variables and operators to their registers and functional units as generated in Figure 6.10 and Figure 6.13, produces the datapath schematic as shown in Figure 6.15. Note that this datapath design requires only seven selector inputs, in comparison with the nine or more selector inputs required by the other design solutions, which did not take merging priorities into account.

6.6   CONNECTION SHARING

In previous sections we have seen how to merge variables and operators and assign them to registers and functional units. After assigning them, however, we still need to connect these registers and functional units into a datapath, wiring each register output to the input of a functional unit and each functional-unit output to the input of a register. The outputs of registers and functional units are called connection sources, and their inputs are called connection destinations. Since several connections can have the same destination at different times, a datapath often includes selectors that are designed to provide the proper connection at the proper time.

Since the connections of a datapath usually occupy a substantial silicon area, we generally try to reduce the number of connections by merging several connections into a bus, which occupies less area. As was the case when we were merging variables and operators, we do this by grouping all those connections that are not being used at the same time and assigning each of these groups to a bus. Each connection source in the group is connected to a bus through a tri-state bus driver which drives the bus in those states in which that source sends data to its destination; otherwise, the source is disconnected from the bus.

The technique for merging connections is similar to those techniques we used for merging variables and operators. First, we create a connection usage table, which indicates the states in which each connection is to be used. Second, we create from this usage table a compatibility graph in which each connection is represented by a node and any two nodes can be connected by a priority edge or an incompatibility edge. As the name implies, two nodes are connected by an incompatibility edge whenever their corresponding connections do not originate from the same source but are to be used at the same time. Conversely, the nodes are connected by priority edges whenever their corresponding connections have a common source or a common destination. Once we have constructed this compatibility graph, we use a graph-partitioning algorithm to group connections in a way that will maximize the number of priority edges included in all groups.

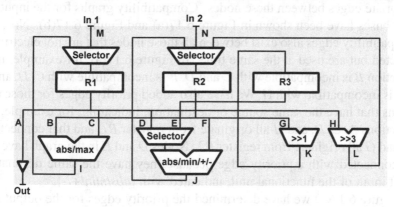

FIGURE 6.16 SRA Datapath with labeled connections

In Figure 6.16, we demonstrate connection merging for the SRA datapath presented in Figure 6.15. Every point-to-point connection is indicated with a letter. From Figure 6.16 and the FSMD model, we can create a connection usage, shown in Table 6.6. In this table, an X has been used to designate the state in which each connection is to be used. Note that this table contains both input connections, which link register outputs to functional unit inputs, and output connections, which link functional unit outputs to the appropriate register inputs. To simplify the partitioning task, it is useful to separate these two types of connections and partition each type separately. By separating these two types of connections, we create separate input and output buses, which will simplify the datapath architecture.

TABLE 6.6 Connection usage table

	S0	S1	S2	S3	S4	S5	S6	S7
A								X
B			X				X	
C		X	X				X	
D			X		X			
E						X		
F		X	X		X	X		
G				X				
H				X				
I		X	X				X	
J		X	X		X	X		
K				X				
L				X				
M	X							
N	X							

Once we have completed the usage table, we can then transform it into compatibility graphs by assigning one node for each connection and adding the appropriate edges between these nodes. Compatibility graphs for the input and output buses have been shown in Figure 6.17(a) and Figure 6.17(b). Note that incompatibility edges also exist between all those nodes that are not electrically connected but are used at the same time. In Figure 6.17(a), for example, input connection B is incompatible with C and D, F is incompatible with C, D, and E, and G is incompatible with H. We have also added priority edges for those connections that have the same source or destination, indicating, for example, that connections A, C, D, and H all originate from register R1, and that connections B, F, and G all originate from register R2. Nodes D and E in the graph have also been connected with a priority edge because they have the same destination, the left input of the functional unit, indicated with [*abs/min/+/-*].

In Figure 6.17(b) we have determined the priority edges for the output connections, by proceeding in a similar fashion. At this point, after we have determined all the priority and incompatibility edges, we can partition the connections, trying to cut all the incompatibility edges while cutting as few priority edges as possible. As shown in Figure 6.17(a), the fewest possible partitions can be achieved by grouping connections A, C, D, E, and H into *Bus1*, and connections B, F, and G into *Bus2*, which accounts for all of the input connections. Similarly, we group I, K, and M into *Bus3* and J, L, and N into *Bus4*, which merges all the output connections. In Figure 6.18, you can then see that the SRA datapath can be connected with a total of four buses, which substantially reduces the cost of connectivity implementation.

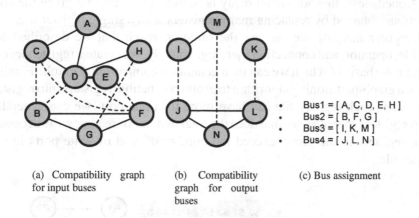

(a) Compatibility graph for input buses

(b) Compatibility graph for output buses

(c) Bus assignment

- Bus1 = [A, C, D, E, H]
- Bus2 = [B, F, G]
- Bus3 = [I, K, M]
- Bus4 = [J, L, N]

FIGURE 6.17 Connection merging for SRA

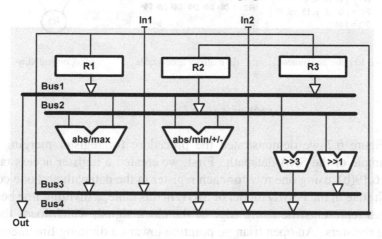

FIGURE 6.18 SRA Datapath after connection merging

6.7 REGISTER MERGING

In Section 6.4 we described a procedure for variable merging which resulted in several variables sharing the same register. As we explained, a number of variables share the same register whenever they have non-overlapping lifetimes. In the same fashion, registers with non-overlapping access times can be merged into register files to share the register input and output ports, which in turn reduces the number of connections in the datapath, because there will be fewer ports. Unfortunately, it also increases the register-to-register delay because an extra delay is incurred for the address decoding that occurs in the register file. Nonetheless, this additional delay is frequently acceptable given the cost reductions obtained by replacing many registers with a single register file.

In register merging we can use the same approach that we described for variable, operator, and connection merging. Initially, we create a register access table, on the basis of which we can then generate a compatibility graph. Finally, we use a graph-partitioning algorithm to group compatible registers into register files. Since each register file can have more than one port, we can generally group registers so that at no time does the total number of read or write accesses to the registers in the group exceed the number of read or write ports in the register file.

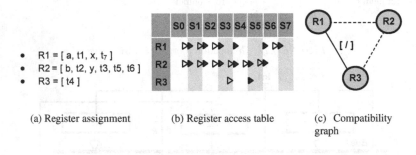

| | (a) Register assignment | (b) Register access table | (c) Compatibility graph |

FIGURE 6.19   Register merging

In Figure 6.7 we demonstrated the procedure for register merging using the example of the SRA datapath. First, we created a register access table in Figure 6.19(b), using one row for each register in the datapath and one column for each state in the FSMD model of SRA. In this table, a dividing line between the states represents the rising edge of the clock signal, which loads the data into the registers. An open triangle pointing toward a dividing line means that new data will be written into the register at that particular rising edge of the clock signal. We have also drawn a black triangle pointing away from a dividing line when we need to indicate the state in which the data will be read from the register file.

From the register access table we can then generate a compatibility graph. In the case of the SRA datapath, we can see that registers *R1* and *R2* are not compatible because they are written or read concurrently in states *S0* through *S4*, and *S6*. Similarly, *R2* and *R3* are not compatible because both are written in state *S3* and read in state *S5*. On the other hand, registers *R1* and *R3* are compatible simply because they are never accessed at the same time. These conclusions are reflected in the compatibility graph shown in Figure 6.19(c), which shows that we can merge registers *R1* and *R3* into a single register file with one read and one write port. The final datapath using such a register file is shown in Figure 6.20.

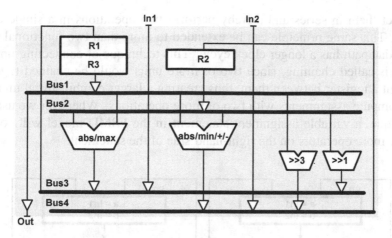

FIGURE 6.20 Datapath schematic after register merging

From this schematic we can also see that by merging registers *R1* and *R3*, we have been able to reduce the number of bus drivers in the datapath because *R1* and *R3* share the same read port, so we need only one bus driver instead of two. In general, merging n registers that drive m buses into a single register file with one read port will reduce the number of drivers by n - m. However, if we merge n registers that are loaded from m different buses into a single register file with one input port, we have to introduce an m-input selector in front of the input port. Because of potential savings in bus drivers and input selectors, the priority in merging registers is generally given to registers with a common source or destination, that is, to registers that are loaded from the same bus or that drive the same bus.

6.8   CHAINING AND MULTI-CYCLING

So far, we presented techniques for datapath synthesis that are based on a simple datapath model. For example, in these datapaths, the registers were connected by one or more buses to the functional units, and the functional units in turn were connected by one or more buses to the registers. In some cases, selectors were used whenever a register or a functional unit received data from more than one bus. In this kind of datapath, the registers are clocked by a clock signal whose cycle is equal to the worst register-to-register delay. Since the worst register-to-register delay path goes through the slowest functional unit, this means that other functional units are busy for only part of the clock cycle and remain idle for the rest of the cycle. If, however, the total delay of any two of these functional units is shorter than the clock cycle, it is possible to connect them in series and thereby perform two operations in a single clock cycle. This same principle can be extended to more than two functional units if the datapath has a longer clock cycle. This technique of connecting units in series is called chaining, since two or more units would be chained together without a register between them, thus creating a larger combinatorial unit that can compute assignments with two or more operations. Whenever we use this technique, a variable assignment statement in the FSMD model will contain two or more operators on the right-hand side of the statement.

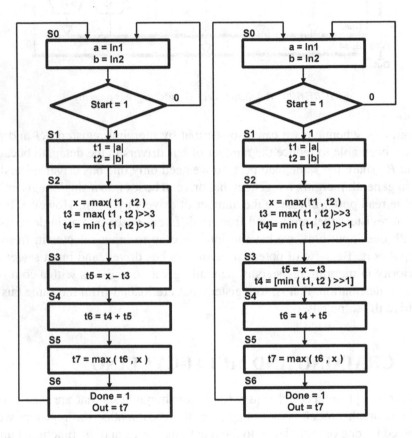

(a) FSMD model for functional unit chaining

(b) FSMD model for functional unit multi-cycling

FIGURE 6.21   Modified FSMD models for SRA algorithm

To demonstrate chaining, we use a modified FSMD model of SRA algorithm, as shown in Figure 6.21(a). Note that this model merges two states (*S2* and *S3*) from the previous FSMD model into one state (*S2*). As you can see, this means that three assignment statements will be executed in state *S2* of Figure 6.21(a);

the first of these statements requires one binary operation (maximum), while the other two statements require two operations each. More specifically, the new value would be assigned to *t3* by computing the maximum of *t1* and *t2* and then shifting the result to the right by three positions. At the same time, the new value for *t4* will be obtained by computing the minimum of *t1* and *t2* and then shifting the result one position to the right. Since shifting to the right by three or one positions incurs no delay, the clock cycle for this chained datapath would be no longer than the original clock cycle. On the other hand, since this FSMD model has only seven states instead of the eight states in the original model, we would conclude that this modified datapath can perform the SRA algorithm 12.5% faster.

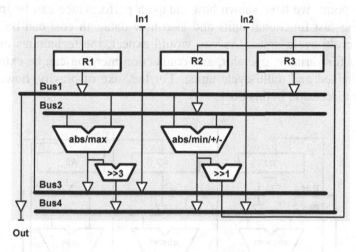

FIGURE 6.22 Datapath with chained functional units

The new datapath schematic with the chained units is shown in Figure 6.22. Note that we had to create an additional connection from the right shifter to register *R3*, so as to concurrently store the new values for variables *x*, *t3*, and *t4* that were generated in state *S2*. Though chaining allows us to concatenate faster units, there are instances in which we must use units which are slower, taking more than one clock cycle to generate results, but which are less expensive. This technique is called multi-cycling, and these slower units are called multi-cycle units. For obvious reasons, such units can be used only for the non-critical paths through the FSMD model. For example, in Figure 6.21(a), variable *t4* will be assigned a new value (min (t1, t2) >> 1) in state *S2*, but this new value will not be used until state *S4*. In this case, then, we could use a unit that takes two clock cycles to compute the minimum value, and chain this unit with a right shifter that takes no time to generate its result.

Such a multi-cycling arrangement is shown in Figure 6.21(b), in which the FSMD model for the SRA has been modified by the introduction of square brackets, used to indicate that the result will only be available in some successor state or that the computation of an expression was already started in one of the predecessor states. For example, the variable assignment [t4] = (min(t1, t2)) >> 1 indicates that the new value will be assigned to t4 in one of the successor states. Similarly, the expression t4 = [(min(t1, t2)) >> 1] indicates that the new value is assigned to *t4* in the present state but that computation of the expression in brackets has started in one of the previous states. As illustrated in Figure 6.21(b), such an FSMD model is easily translated into a datapath with multi-cycle units. Such multi-cycle units show up in many datapaths with fast clock cycle, in which some functional units need two or more cycles to finish their operation.

At this point, we have shown how datapath performance can be improved by chaining fast functional units and also how datapath cost can be reduced by using multi-cycle units. As you would expect, the techniques described previously for variable, operator, and connection merging can be extended to include chained and multi-cycle units. For the sake of brevity, however, we omit their discussion in this book.

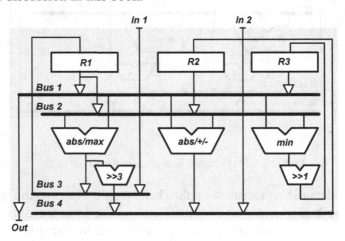

FIGURE 6.23   SRA datapath with chained and multi-cycle functional units

6.9   FUNCTIONAL-UNIT PIPELINING

In previous sections, we introduced various techniques for reducing datapath cost, mainly by reducing the number of registers, functional units, and connections. In this section, we shift focus by introducing techniques that increase the performance of a datapath. The single most effective technique for performance improvement is pipelining. Pipelining can be applied to functional units, datapaths, and controllers.

To pipeline functional units, we divide a functional unit into two or more stages, each separated by latches so that each stage can operate on a different set of operands. At any time, then, there are several sets of operands in the pipeline. More precisely, the number of sets in a pipeline equals the number of its stages. Using pipelined functional units does not affect the time taken to generate results for the first set of operands, which is approximately the same as the time in a non-pipelined unit. However, for every additional set of operands, a result is available in a time equal to the delay of only one stage. For example, for a 2-stage pipelined unit, whose non-pipelined delay is 10 ns, the result for the first set of operands is still generated in 10 ns, but the result for the second set of operands is available only 5 ns later, as will every subsequent result in that pipeline . In general, if there are n stages in the pipeline, we can reduce the time taken to generate results to approximately 1/n times the non-pipelined execution time, with the exception of the first result. Unfortunately, functional units can not be easily divided into n stages of equal delay. Usually different pipeline stages have slightly different delays. Thus, the clock cycle of the pipelined unit is the longest of all the pipelined stages delay, which results in a clock cycle that is larger then 1/n of the clock cycle of the original non-pipelined unit.

To compare the results of pipelined and non-pipelined units, let us consider an SRA datapath with only one non-pipelined ALU, which performs absolute value, minimum, maximum, sum, and difference according to the FSMD model shown in Figure 6.24(a). Note that this datapath requires nine states, or nine clock cycles, to compute a square-root approximation. On the other hand, we could redesign the datapath by replacing its non-pipelined ALU with a 2-stage pipelined ALU, as shown in Figure 6.24(b). This new datapath with the pipelined ALU requires 13 clock cycles to compute the square-root approximation, as shown in the timing diagram in Figure 6.24(c). However, pipelined-datapath clock cycle will have approximately half the duration of the non-pipelined clock cycle. Therefore, the pipelined design finishes the SRA faster then non-pipelined design.

In the timing diagram in Figure 6.24(c), the reading and writing of each register and the operation of each functional unit are shown on a clock-by-clock basis. The timing diagram has one row for each register read or write, as well as one row for each stage of the ALU and shift units. Each column represents one clock cycle. As you can see, in the clock cycle corresponding to state $S0$ in the FSMD representation, the Datapath reads the values of variables a and b from the input ports and stores them in registers $R1$ and $R2$. In the next clock cycle, the Datapath executes state $S1$ in which it reads the value of variable a from register $R1$ and partially computes the absolute value of a. This partial result is stored in the ALU pipeline latches between two stages.

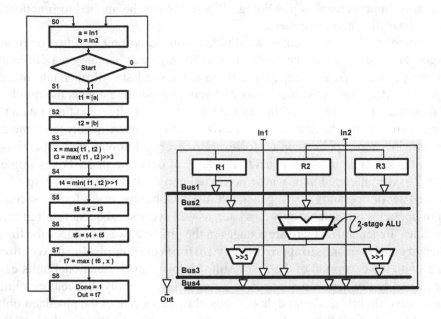

(a) FSMD model   (b) Datapath with 2-stage pipelined ALU

	S0	S1	S2	NO	S3	S4	S5	NO	S6	NO	S7	NO	S8
Read R1		a			t1	t1	X				X		t7
Read R2			b		t2	t2	t3		t5		t6		
Read R3									t4				
ALU stage 1		|a|	|b|		max	min	-		+		max		
ALU stage 2						max	min	-		+		max	
Shifters							>>3	>>1					
Write R1	a		t1			X						t7	
Write R2	b				t2		t3		t5		T6		
Write R3									t4				
Write Out													t7

(c) Timing diagram

FIGURE 6.24 Functional unit pipelining

Then, in the next clock cycle, the Datapath finishes the computation of |*a*| and assigns this value to variable *t1*, which is stored in register *R1*. At the same time, the Datapath also initiates the computation of |*b*| specified in state *S2*, and storing the partial result of this computation in the pipeline latches. Thus in this clock cycle both stages of the pipelined ALU are active, although they process different operands. In next clock cycle, the Datapath finishes computation of |*b*| and assigns it to variable *t2* stored in register *R2*. Note that at this point the

Datapath cannot yet initiate the next operation because it requires the value of *t2*, which has not yet been loaded in register *R2*. Therefore, no operation is scheduled to start in fourth clock cycle and operation specified in state *S3* is delayed by one clock cycle. In a similar fashion, the Datapath starts execution of maximum, minimum, and subtraction operations in the fifth, sixth, and seventh clock cycles, and completes these operations, together with the shifts, in clock cycles six, seven, and eight. It cannot start the addition specified in state *S7*, since it must wait for the availability of the value assigned to *t5*. Similarly, it starts the maximum operation in eleventh clock cycle with result being written into register *R1* in the next clock cycle. Finally, the Datapath uses the thirteenth clock cycle to output the result.

According to this timing diagram, the SRA algorithm requires 13 clock cycles to complete. As mentioned above, however, two of these clock cycles are equal to one clock cycle of the non-pipelined design, which means that the datapath with the pipelined ALU computes the square-root approximation in six and a half clock cycles instead of the nine needed by the non-pipelined design, which is 28% faster than the non-pipelined design. Note that this pipelined datapath can outperform any non-pipelined design described in previous sections.

6.10   DATAPATH PIPELINING

Just as with a functional unit, we can pipeline the whole datapath by inserting registers in some coordinated fashion on every register-to-register path. The best way to pipeline a datapath is to divide the register-to-register delay into stages, which is easiest to achieve by inserting registers at the inputs and outputs of the functional units. A pipelined datapath for the SRA computation is shown in Figure 6.25(a). The register-to-register path is divided into three stages: registers to ALU input, ALU input to ALU output, and ALU output back to registers. Note that ALU by itself can be also pipelined as described in previous section.

In the timing diagram in Figure 6.25(b), the reading and writing of each register and the operation of each functional unit are shown on a clock-by-clock basis. The timing diagram has one row for each register read or write, as well as one row for the ALU input and output registers and shift units. Each column represents one clock cycle where the column headings indicate the state of the FSMD model that is initiated in that clock cycle or if no state is initiated in that clock cycle. As you can see, in the first clock cycle corresponding to state *S0* in the FSMD representation in Figure 6.24(a), the datapath reads the values of variables *a* and *b* from the input ports and stores them in registers *R1* and *R2*. In the next clock cycle, the datapath executes state *S1*, in which it reads the value of variable a from register *R1* and stores it in the ALU input

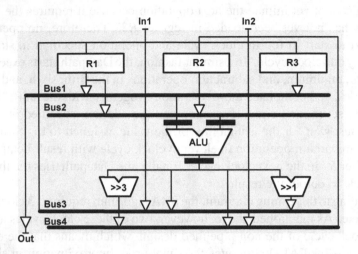

(a) Pipelined datapath for SRA computation

	1	2	3	4	5	6	7	8	9	10	11	12	13	14	15	16	17	18
Read R1		a				t1	t1	x							x			t7
Read R2			b			t2	t2		t3			t5			t6			
Read R3												t4						
ALUIn(L)		a				t1	t1	x				t4			x			
ALUIn(R)			b			t2	t2		t3			t5			t6			
ALUOut			|a|	|b|			max	min		-				+		max		
Shifters								>>3	>>1									
Write R1	a			t1				x									t7	
Write R2	b				t2				t3			t5			t6			
Write R3									t4									
Write Out																		t7

(b) Timing diagram

FIGURE 6.25 Datapath pipelining

register. In the next clock cycle, it computes the absolute value of *a* and stores
it in the ALU output register. Then, in the following clock cycle, the datapath
finishes the computation of |*a*| and assigns this value to variable *t1*, which is
stored in register *R1*. In the third clock cycle, the datapath also initiates the
computation of |*b*| specified in state *S2* and stores *b* in the ALU input register.
In the subsequent clock cycle it computes |*b*| and stores it in the ALU output
register. Finally, in the next clock cycle, it moves |*b*| to register *R2*. Note that
in the fourth and fifth clock cycle the datapath cannot initiate state *S3* because
it requires the values of *t1* and *t2*, which have not yet been loaded into registers

R1 and *R2*. Therefore, no operation is scheduled to start in the fourth or fifth clock cycles. Thus, the operations specified in state *S3* are delayed by two clock cycles to start in the sixth clock cycle and finish in the eight clock cycle. In a similar fashion, the datapath starts execution the state *S4* in seventh clock cycles and completes it, together with the shift operation, in the ninth clock cycle. It can start the subtraction operation specified in state *S5*, in the same clock cycle since the values of variables to *x* and *t3* are available in registers *R1* and *R3*.

From the above description we can see, that this datapath needs three clock cycles to finish any operation specified in any state of the FSMD model and may require two no-operation (NO) cycles if there is a data dependency between two operations in the two succeeding states in the FSMD definition. In other words, it requires two NO cycles if a variable is assigned a value in one state of the FSMD description and used in the next state of the same description.

According to the timing diagram in Figure 6.25(b), this SRA algorithm requires 18 clock cycles to complete. As mentioned above, however, three of these clock cycles are approximately equal to one clock cycle of the non-pipelined design, which means that the pipelined datapath in Figure 6.25(a) computes the square-root approximation in six clock cycles instead of the nine needed by the non-pipelined design, so it arrives at the result 30% faster than a non-pipelined SRA data path. Note that datapath pipelining described in this section can outperform functional unit pipelining that has been described in the previous section. In general, combining both types of pipelining may result in greater performance improvement than for each type of pipelining by itself. However, adding more pipeline stages may not be profitable in case of the SRA example since SRA does not have enough operations that can be executed concurrently. In general, the number of pipeline stages is equal to the number of operations that can be executed concurrently. If there is not enough concurrency, the pipeline has to wait for data to become available by performing no-operation cycles.

6.11 CONTROL AND DATAPATH PIPELINING

In previous sections, we discussed two methods for improving performance through pipelining techniques that reduce the register-to-register delay. It is important to note, however, that the longest register-to-register delay can be usually found in the control unit. This critical path through the controller determines the length of the clock cycle or clock period. Consequently, if we want to improve performance by shortening the clock cycle, it would make sense to divide the critical path through the controller into pieces and insert registers between them. In Figure 6.26(b), for example, registers are inserted in three difference places. First, we introduce a Status Register (SR) between the datapath and the controller so that all status signals leaving the datapath are latched in that register, which has one flip-flop for each status signal. Second, we insert a Control-Word Register (CWR) between the control unit and the Datapath so that all control signals generated by reading the CMem are stored in that register. Finally, we pipeline the datapath itself by inserting pipeline registers between the storage units (register, RF, and memory) and the functional units (ALU, multiplier/divider), and between functional units and storage units again, as explained in previous sections. This way it takes four clocks from PC to PC and from PC to any storage unit in the datapath.

In general, when we plan to use control and datapath pipelining, we may need to wait for a clock cycle or two if control words or data is not available. To demonstrate this, let us consider a small part of a FSMD model shown in Figure 6.26(a), which in its original form has three states. In the first state, $S1$, we test whether $a > b$, then go to $S2$ if this inequality is not true or to $S3$ if it is true. In state $S2$, we execute the assignment $x = c * d$, and in state $S3$ we would execute the assignment $y = x - 1$. Note that this FSMD model does not assume any pipelining in its definition.

The timing diagram in Figure 6.26(c) shows the execution of the FSMD model in Figure 6.26(a) on the design in Figure 6.26(b). We will assume that the control word for $S1$ is stored in the CMem at address 10. In clock cycle #1, it is in the CWR so that data a and b are fetched from RF and stored in the ALU Input registers. Then in clock cycle #2, a and b are compared in the ALU and written into the SR. In the next clock cycle, CMem address 14 or 17 is written into the PC depending on whether $a > b$ or not. If $a > b$, then the Datapath executes the control word written at location 14 in the CMem. Otherwise, it jumps to location 17 and writes the control word written there to CWR. In case the datapath executes $S2$ the control word for $S2$ will be in the CWR in clock cycle #5. In that clock cycle, the datapath fetches variables c and d from the RF and stores them in the ALU input registers. In clock cycle #6, the datapath computes $c + d$ and stores the result in variable x in the RF during clock cycle #7. Assignment of the variable $y = x - 1$ cannot start until after clock cycle #7, when x is written into the RF. Therefore, the execution of $y = x - 1$ starts in clock cycle #8, when x and 1 are fetched from RF and stored in ALU input registers. The computation of $x - 1$ is performed in clock cycle #9, while storing the result in variable y in the RF is accomplished in clock cycle #10.

Note that states $S1$, $S2$, and $S3$ in the FSMD model depend on each other. Therefore, the execution of each state can start when the address for the next control word has been computed and loaded into the PC or when the variable value computed in the previous state is in the RF. The control words for $S1$, $S2$, and $S3$ are stored in the CMem at locations 10, 14, and 17. Other locations store no-operation (NO) control words. Therefore, after $S1$ is loaded into the CWR in clock cycle 0, it takes three clock cycles until a new address (14 or

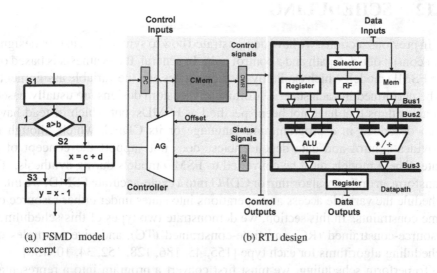

(a) FSMD model excerpt   (b) RTL design

Clock cycle #	0	1	2	3	4	5	6	7	8	9	10
Read PC	10	11	12	13	14	15	16	17	18	19	20
Read CWR		S1	NO	NO	NO	S2	NO	NO	S3		
Read RF(L)		a				c			x		
Read RF(R)		b				d			1		
Write ALUIn(L)			a				c			x	
Write ALUIn(R)			b				d			1	
Write ALUOut							c+d			x-1	
Write RF								x			y
Write SR			a>b								
Write PC	11	12	13	14/16	15	16	17	18	19	20	

(c) Timing diagram

FIGURE 6.26 Control and datapath pipelining

17) is loaded in the PC and one more clock cycle until the new control word is loaded into the CWR. Therefore, the datapath executes three NO control words stored at locations 11, 12, and 13 as it waits for $a > b$ comparison to complete. Similarly, datapath, while waiting for variable x to be stored in *RF*, executes two NO control words stored at locations 15 and 16 in the CMem.

From this simple example, we can see that control and datapath pipelining does not help if there is control dependency (*S2* depends on *S1*) or data dependency (*S3* depends on *S2*). However, pipelining does help if there are at least three not dependent states in the FSMD model or three independent statements in a BB. Control dependency is difficult to avoid since the next address is computed dynamically. But it is possible to minimize the number of NO words in a datapath with branch prediction, as long as the prediction is correct most of the times.

6.12   SCHEDULING

In previous sections, we have demonstrated how to synthesize custom designs that consist of a datapath and a control unit. In general, the synthesis is based on the FSMD model, which explicitly specifies states and the variable assignments to be performed in each state. Unfortunately, custom designs are usually based on algorithms that have not been specified as FSMDs, but which instead have been described in a programming language or its CDFG, which, though it provides control and data dependencies, does not support the concept of a state. These models must be converted to FSMD models during synthesis. To transform an ordinary algorithm or CDFG into a cycle-accurate FSMD, we must schedule the variable access and operations into states under either resource or time constraints. In this section, we demonstrate two types of this scheduling, resource-constrained (RC) and time-constrained (TC), and give examples of scheduling algorithms for each type [155, 45, 186, 128, 152, 34, 10, 156].

To perform scheduling, we must first convert a program into a representation, such as CDFG, which explicitly shows the control dependencies among statements as well as the data dependencies among variable values. In Figure 6.26(a) and Figure 6.26(b), we show a C-language flowchart and its corresponding CDFG for the SRA algorithm. Such a CDFG representation is frequently used by scheduling algorithms, since a scheduled CDFG is equivalent to a cycle-accurate FSMD. As we mentioned earlier, a CDFG can be scheduled using resource or timing constraints. For the former, we can specify the resource constraints by the complete or partial number and type of functional and storage units, and their connections, in the datapath. They can also be specified by giving a complete netlist of the datapath. For scheduling using timing constraints, the time constraints are specified as the number of states or clock cycles the datapath will need to execute all the operations on the longest path through the given CDFG.

However, before we detail RC and TC scheduling algorithms, we must introduce as-soon-as-possible (ASAP) and as-late-as- possible (ALAP) scheduling algorithms, which are frequently used by RC and TC scheduling algorithms to determine operation priority and range for scheduling.

ASAP and ALAP algorithms assume, first, that each operation will take exactly one clock cycle to execute, and second, that an unlimited number of functional units or resources are available for each operation in each state. Because of these assumptions, both algorithms are constrained only by data dependencies. Within this context, the ASAP algorithm schedules each operation into the earliest state in which all its operands are available. In other words, it scans the CDFG from top to bottom and assigns to each state all the nodes in the graph whose predecessor or parent nodes have been already assigned into

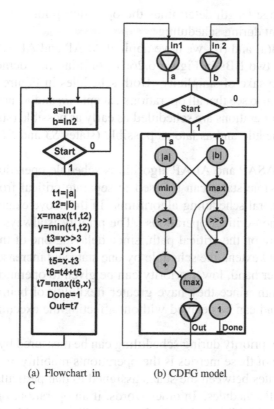

(a) Flowchart in C   (b) CDFG model

FIGURE 6.27   C and CDFG

previous states. Thus the ASAP algorithm generates a schedule that has the minimum number of states or, in other words, the shortest execution time.

In contrast to the ASAP approach, the ALAP algorithm schedules each operation into the last possible state before its result is needed, the ALAP algorithm, using the length of the final schedule in the number of states as a constraint, schedules each operation into the last possible state before its result is needed. In other words, it scans the CDFG from the bottom to the top and assigns to each state all the nodes whose successor or children nodes have been already assigned into later states. If the required schedule length is equal to the length obtained by the ASAP algorithm, we can observe, that the ALAP algorithm schedules all the operations on the critical path through the dataflow graph into the same states as the ASAP algorithm. The operations that are not on the critical path are scheduled earlier than needed in the ASAP schedule and later than possible in the ALAP schedule. Therefore, ASAP and ALAP algorithms give us ranges of possible states for scheduling non-critical operations. Furthermore, the range length determines the operation priority or urgency to be taken into account during scheduling.

In Figure 6.28(a) and (b), we have applied ASAP and ALAP scheduling to the larger of the two BBs in Figure 6.26(b), limiting this demonstration to a single BB for the sake of simplicity. Both schedules in Figure 6.12.1 require seven states. We also see that all operations except *min* and *»1* are on the critical path. These two operations are scheduled as early as possible (states $S2$ and $S3$) in the ASAP schedule and as late as possible (states $S3$ and $S4$) in the ALAP schedule.

The fact that ASAP and ALAP algorithms schedule operations on the critical path to the same states can be used to separate critical from non-critical operations in general scheduling algorithms. In the above example, therefore, *min* and *»1* are non-critical operations. The priority is always given to those operations that are on the critical path, since delaying one of these operations by one state would extend the schedule by one state and increase the execution time. On the other hand, lower priority can be given to operations that are not on the critical path, since they have greater flexibility of being scheduled in different states and can be delayed without affecting the execution time of the entire CDFG.

An operation's priority during scheduling can be measured by several different metrics. One of these metrics is the operation's mobility, which is equal to the number of states between the states assigned to that particular operation in ASAP and ALAP schedules. In other words, if an operation, *op*, is scheduled in state Si in the ASAP schedule and in state Sk in the ALAP schedule, its mobility, $M(op)$, will be equal to $k - i$. Thus mobility defines the operation's ability to be postponed without an impact on the total execution time. This can be used for prioritizing operations because states with higher mobility can be given lower priority.

As an alternative measure of priority, we can use the criterion of operation urgency, which is equal to the distance in the number of states between the state in which the operation is available for scheduling and the state in which the operation occurs in its ALAP schedule. In other words, if an operation, *op*, is available in state Sj but is not scheduled until state Sk in its ALAP schedule, that state's urgency $U(op)$ will be equal to $k - j$.

As a third measure of operation's priority, we could consider how many other operations use its result as an operand, or in other words, the number of dependencies. This measure gives priority to operations that increase the number of operations available for scheduling in the future. There are several other priority measures, but none of them works perfectly in all cases. In theory, we could use any number of priority metrics in any order to determine a priority for scheduling among several operations available at the same time.

6.12.1 RC SCHEDULING

One of the most popular algorithms for RC scheduling is the list-scheduling algorithm, which uses a ready list of operations that are available for scheduling. In this algorithm, the operations on the ready list are sorted by their mobilities, so that the operations with zero mobility will be placed at the top of the list while those operations with the greatest mobility will be placed at the bottom. In cases where two operations have the same mobility, priority is given to the operation with the lower urgency number. If those are identical as well, the priority is assigned randomly. In applying this list-scheduling algorithm, we take the following steps in each state: assign the highest-priority operations from the ready list to the available functional units, one at a time, then delete all the assigned operations from the list, and insert the newly schedulable operations into the list in the positions that correspond to their mobilities and urgencies.

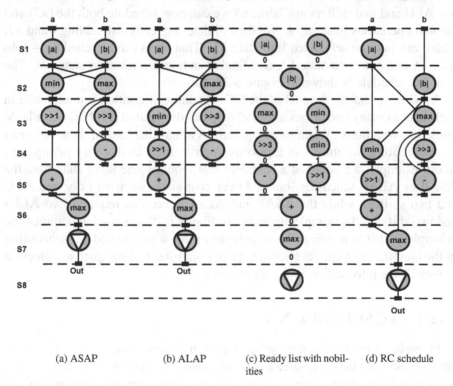

(a) ASAP   (b) ALAP   (c) Ready list with nobil-   (d) RC schedule
                                        ities

FIGURE 6.28   ASAP, ALAP, and RC schedules for SRA

This list-scheduling algorithm is summarized in FIGURE 6.29. We demonstrate this list-scheduling algorithm on the dataflow graph of the BB in Figure 6.27(b) under the assumption that we have only one arithmetic unit, which can perform absolute value, minimum, maximum, addition, and subtraction, in addition to two shift units. To perform RC scheduling with a list-scheduling algorithm, we first generate ASAP and ALAP schedules, as shown in Figure 6.28(a) and (b). Then we create a ready list for the first state, as shown in Figure 6.28(c), and compute the mobilities for the operations in that ready list. In our case, only operations |a| and |b| are available in the first state. Since these operations have the same mobility [$M(|a|) = M(|b|) = 0$], we select their order randomly and schedule |a| first. Since scheduling |a| does not free any more operations for scheduling, we do not change the ready list at this time. Therefore, we must schedule |b| in state S2, which allows us to add *max* and *min* operators to the ready list. Since the *max* operator is on the critical path, it has a mobility $M(max) = 0$, which gives it priority over the *min* operator, which has mobility $M(min) = 1$. Therefore, we would schedule *max* into state S3, which allows us to add »3 to the ready list. Since its mobility $M(»3) = 0$, this operator should be placed at the top of the ready list. At this point we have one ALU and two shifters available, so we can now schedule both the (»3) and the *min* operations in state S4. We then update the ready list, adding - and »1, which can both be scheduled into state S5. That allows us to schedule + into state S6 and *max* into state S7, and finally, output the result in state S8. The final RC schedule is shown in Figure 6.28(d).

As we have shown, the goal of the RC scheduling algorithm is to schedule in each state as many operations as possible given the limited number of available units or connections. When more operations are available than there are units or connections, we must use a priority metric, such as mobility or urgency. In our example, we obtained a schedule that requires one more state than the ASAP or ALAP schedules, but at lower cost since we used only one ALU and two shifters, while the ASAP and ALAP schedules required two ALUs and two shifters. The simple algorithm in Figure 6.29 was used to demonstrate principles of list scheduling. There are many more sophisticated list-scheduling in the literature that can take advantage of controller and datapath architecture as well as the information in the input model.

6.12.2   TC SCHEDULING

In many cases, the primary goal of design optimization is to improve the performance, not the cost, since a datapath must execute a given code in a fixed amount of time. When execution time is our priority, we use time-constrained (TC) scheduling, which generates a schedule comprising a particular number of states while attempting at the same time to minimize the number of functional units it requires in the datapath. This goal is achieved by creating a probability-distribution graph and using it to schedule operations into states one at the time so that the largest sum of probabilities for each operator and each state is minimal.

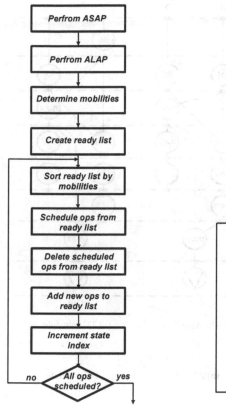

FIGURE 6.29 RC algorithm   FIGURE 6.30 TC algorithm

To use TC scheduling, we first apply the ASAP and ALAP scheduling algorithms to determine the mobility range for each operation. Having established these ranges, we then assign to each operation an equal probability of being scheduled in each state in its range. Obviously, for each operation, the sum of all these probabilities over the entire range equals 1.

Once we have calculated these probabilities, we can then create probability distribution graphs, which define the probability sums in each state for each set of compatible operations. In other words, these probability sums determine the number of functional units of each type required in each state. Using this information, we can attempt to minimize the number of functional units by selecting an operation and scheduling it in the state that will reduce the largest probability sum for this operation type in the distribution graph. If reduction is not possible, we can select an operation and schedule it in the state in which it will minimally increase the probability sum. The algorithm terminates only when all operations have been scheduled as shown in Figure 6.30.

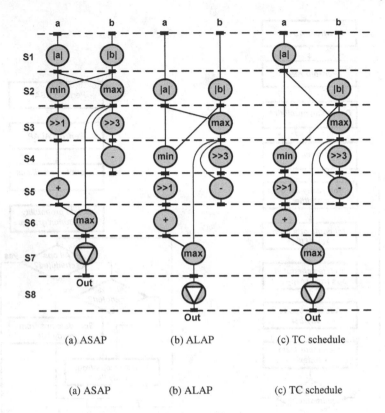

(a) ASAP   (b) ALAP   (c) TC schedule

FIGURE 6.31   ASAP, ALAP, and RC schedules for SRA

We demonstrate TC scheduling on the dataflow graph of the larger BB in Figure 6.27(b), and for comparison purposes we set a goal of eight states for the complete schedule, as this was the schedule length that we obtained with RC scheduling using one ALU and two shift units. In our first step, we create ASAP and ALAP schedules, as shown in Figure 6.31(a) and (b).

From these schedules, we can compute mobility ranges, concluding that the mobility range for all the operators except *»1* and *min* would be 2. In other words, the probability of each of these operators being scheduled in any particular state in its range would be 0.5. Since the operators *min* and *»1* each have a range of three states, the probability of their being scheduled in any particular states within that range would be 0.33. These individual probabilities are combined in Figure 6.32(a) into two distribution graphs that we then use for minimizing the number of ALUs and shift units.

As you can see from the distribution graphs for ALUs and shift units, scheduling any operation into a particular state increases the probability sum in that state and therefore the number of units required, except in state *S7*, where the

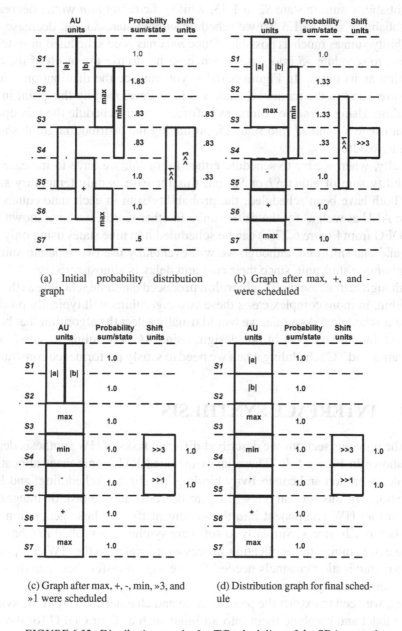

(a) Initial probability distribution graph

(b) Graph after max, +, and - were scheduled

(c) Graph after max, +, -, min, »3, and »1 were scheduled

(d) Distribution graph for final schedule

FIGURE 6.32 Distribution graphs for TC scheduling of the SRA example

probability sum is only 0.5. Therefore, we schedule the *max* operator in state *S7*, which increases the probability sum to 1.0. For the same reason, we schedule + in state *S6* and - in state *S5*. At this point we have the option of scheduling *max* or *min* in state *S3*. As you can see, however, scheduling *max* there decreases the probability sum in state *S2* to 1.33, while scheduling *min* would decreases the probability sum to 1.5, so we schedule *max* in state *S3* and decrease this probability sum as much as possible. Once *max* has been scheduled in state *S3*, we have to schedule *»3* into state *S4* since *»3* has to use the result of the *max* operation as its input. In Figure 6.32(b), you can see the distribution graphs that correspond to the partial schedule we have developed by this point in the algorithm. Using the same criteria as before, we then schedule the *min* operation into state *S4* and *»1* into state *S5*, producing the distribution graph shown in Figure 6.32(c).

Finally, when we try to schedule either |a| or |b|, we have to increase the probability sum of either *S1* or *S2*, but this increase is only temporary since when both have been scheduled, the probability sum in each state equals 1.0 for the ALU as well as for the shifter unit. In other words, we have shown that the CDFG from Figure 6.27(b) can be scheduled into nine states using only one ALU and one shift unit, although we will eventually use two separate shifters instead of one shift unit, since their cost and delay is almost zero.

Although in this case the TC algorithm produced the same schedule as the RC algorithm, in more complex cases these two algorithms will typically produce different schedules. As a rule, we would usually select the algorithm that better matches the primary goal of our design, using RC scheduling to satisfy cost constraints and TC scheduling when we need to satisfy performance constraints.

6.13   INTERFACE SYNTHESIS

In the previous sections we described different tasks of HW synthesis design flow shown in Figure 6.1. These tasks included RTL component allocation, variable, operation and connectivity binding, pipelining, scheduling, and RTL generation. In this section we will try to demonstrate the task of integrating such custom HW component into the system platform whose generation was described in Chapter 4. Similarly to software synthesis described in Chapter 5 we have to combine the application processes allocated to the HW component with communication channels needed for message transfers between different components in the platform.

First, we need to extract the process code and channel code from the system TLM model and combine them into an input such as C or CDFG for the HW synthesis tools. However, the synthesized HW component may use different clock cycle then the bus protocol that is used for the communication among components in the platform. To avoid this unusual constraint to HW component synthesis we may want to separate sending a message with bus protocol from computation performed in the HW component. For this reason we need to introduce a special bus interface that sends the message produced by HW component to another platform component with the bus protocol timing.

Second, the main advantage of automatic HW synthesis is its ability to use an untimed code as an input and generate a cycle-accurate RTL description. As a result, the designer is relieved from defining cycle timing. While this scheduling freedom is beneficial for implementing application and some of the communication code it becomes severely limited by imposing protocol timing constraints on synthesis tasks. In order to avoid this constraint and still guarantee bus protocol timing is to not synthesize the bus interface, but to instantiate a pre-implemented and tested bus interfaces from a library. In this case, a special driver layer is used to interface the synthesized application and communication code and the pre-implemented bus interface.

Figure 6.33 outlines an implementation stack for a custom HW component. This implementation stack is partly contained in the system model, which we have described in Chapter 3. The stack generation is similar to the software synthesis described in Chapter 5. The shown HW component includes an application process *P2* which has been combined with a communication stack, that models communication with external components. The stack layers are divided into three groups based on the timing constraints for the generated implementation.

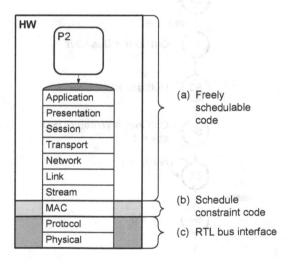

FIGURE 6.33 HW Synthesis timing constraints

(a) **Freely schedulable code**. Most of the code inside a custom hardware component is freely schedulable. This includes the actual application code process *P2* and most of the communication stack down to a set of MAC layer calls for sending and receiving a word or packet on the bus.

(b) **Schedule constraint code**. Schedule constraint code contains timing limitations that have to be observed during synthesis. The MAC driver code is an example of the schedule constraint code. It interfaces between the freely schedulable code, and the pre-implemented bus interface component. Its code is implemented as a function that the freely schedulable code can call in order to drive and sample control and data ports for communication with the bus interface in a cycle defined fashion. The MAC driver is specific for the selected bus interface.

(c) **Bus interface**. The bus interface component that connects custom processor or HW component to a common shared bus possesses the strictest timing constraints. The bus wires have to be sampled according to strict cycle-timing requirements of the bus protocol. To guarantee the specific bus timing, a pre-implemented bus-interface component can be used. It implements the bus protocol state machine using bus clock cycle. The bus interface is usually described in RTL and is inserted from the component library into the RTL description of the entire platform.

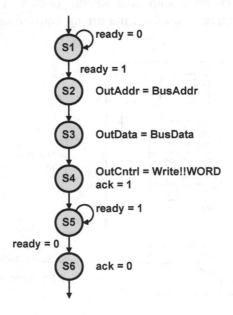

FIGURE 6.34 FSMD for MAC driver

Separating the custom HW implementation into the three categories based on their scheduling constraints, allows us to flexibly synthesize any application code, while still communicating via an arbitrary shared bus with a well-defined bus protocol. The important bridging between the freely schedulable code and the bus interface is the MAC driver. The driver is defined by an FSMD and callable from within the synthesized code. The FSMD defines explicit input and output registers as well as control signals, which are used to interface custom HW and bus interface components.

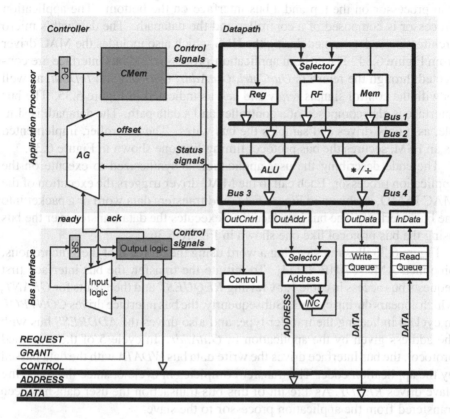

FIGURE 6.35 Custom HW component with bus interface

Figure 6.34 shows a FSMD model of a MAC driver example for communication with a bus interface in a double handshake fashion. It uses output registers *OutAddr*, *OutData* and *OutCntrl*. In addition, it uses control signals *ready* and *ack*. This driver can be invoked from the synthesized communication code as shown in Figure 6.33, by calling the function busif_send(), which then triggers transferring *BusWord* to the address *BusAddr*. The driver first waits until *ready* indicates that the bus interface is free for the next transfer. Then in state *S2*, the driver loads the address register *OutAddr*, in state *S3*the data register *OutData*, and sets the a control register *OutCntrl* for writing a single word on the bus in state *S4*. It also signals by setting *ack* that data is ready for the bus transaction. The driver then waits until the bus interface responds by lowering *ready* that it has received the transfer request. The driver finishes the transaction by lowering *ack* in state *S6*. At this point in time, the bus interface may start the bus transaction.

Figure 6.35 shows a synthesized custom HW component with a application processor on the top and a bus interface on the bottom. The application processor is composed of a controller and the datapath. The datapath's micro architecture is determined during the HLS which also includes the MAC driver from Figure 6.34. The custom application processor and bus interface are connected through the registers *OutCntrl*, *OutAddr*, *OutData* and *InData*, as well as with the control signals *ready* and *ack* as indicated in Figure 6.35. The bus interface is also composed of a controller and a data path. The datapath is simple, as it only drives and samples the bus wires. The controller, implemented as an FSM, secures the bus protocol timing like one shown in Figure 6.36.

The code describing the user application is synthesized to execute on the application processor. Each call to the MAC driver triggers the execution of the MAC FSMD, as shown in Figure 6.34, and transfers data word or a packet into the bus interface. The bus interface then executes the data transfer over the bus using the bus protocol like one shown in Figure 6.36.

Figure 6.36 illustrates writing a word using the protocol of the synchronous, pipelined AMBA AHB 2.0 [4]. To initiate the transfer, the bus interface first requests bus access in cycle 1, by raising *REQUEST* and then waits for *GRANT*, which appears during cycle 3. Subsequently, the bus interface drives *CONTROL* in cycle 4 indicating the transfer type, and also drives the *ADDRESS* bus with the address given by the application in *OutAddr*. In cycle 5 of the pipelined protocol, the bus interface drives the write data bus *WDATA* with the data passed by the application code. The transfer completes in cycle 6, since the receiving slave drives *READY*. As a result of this bus transaction the user data has been transfered from the application processor to the slave.

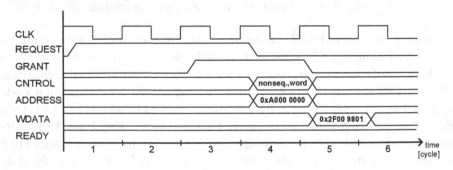

FIGURE 6.36 A typical bus protocol

The above described bus interface solves the problem of transferring data in and out of a synthesized custom HW component. The same solution can be used to transfer data between any two components with different communication protocols. This is particularly true in a complex platform with several busses. In this case we can combine two bus interface components into a bridge that converts one protocol into another. More complex conversion of differently constructed messages can be accomplished with a transducer.

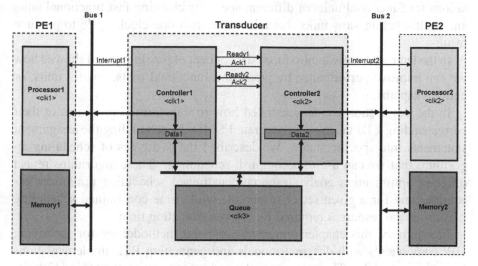

FIGURE 6.37 Transducer structure

Transducer is a device for protocol transformation and routing messages. Messages may have any length and may use any format or protocol known to the transducer. They are stored one message at the time in transducer queue. Each transducer consists of a queue and two message controllers, *Controller1* and *Controller2*, as shown in Figure 6.37. Each controller can write into and read from the queue. Both controllers and the queue may run under different clocks. One controller receives the message with one protocol, decomposes the message into words and stores it in the queue, while the other controller composes message from the queue and sends it out to proper destination under the second protocol. Transducer also computes the new route if it is not encoded in the message.

Therefore, using bus interfaces, bridges and transducers we can construct a multi-bus or network-on-chip communication among variety of components in a multi purpose heterogeneous platform.

6.14 SUMMARY

In this chapter we have explained how to specify and generate HW component design for some standard input models given by CDFG and FSMD models. In addition, we described several procedures for optimizing such designs, showing how to merge variables and assign them to registers, how to merge registers into register files or memories, how to merge operators and assign them to multifunction units, and how to merge connections and create buses for each group of connections. We also demonstrated how to optimize these implementations for functional units of different speed by chaining fast functional units and multi-cycling slow units that take more than one clock cycle to produce results.

In the later sections we introduced the concept of pipelining and showed how we can improve performance by pipelining functional units, control units, or entire datapaths.

In the last section we demonstrated how to transform C programs or their corresponding CDFGs to cycle-accurate FSMDs by scheduling the assignment statements into specific states. We described the two types of scheduling algorithms that we can use to accomplish scheduling. These algorithms reflect different optimization goals: resource-constrained scheduling minimizes execution time for a given set of resources, while time-constrained scheduling minimizes the resources required for a given execution time.

In summary, this chapter presented a general methodology for specifying HW components with different models and generating RTL implementations from these models. There are several good references on the topic [178, 35, 114, 61, 186, 139, 110, 45].

However there is lot of more work needed to improve design quality such as control and datapath pipelining, synthesis optimization for different metrics such as power, manufacturability, dependability, introduction of HLS or architecture cells such as different controllers and datapaths to replace standard cells and work on pre synthesis optimization and input re-coding for synthesis.

Chapter 7

VERIFICATION

Verification is one of the key components of any system design effort. As opposed to device testing, verification involves analysis and reasoning on a computer model of the system before it is manufactured. It is crucial for a designer to ascertain the highest degree of confidence in a product's functional correctness before it is shipped. Economic as well as safety reasons make verification so central to system design. Safety critical systems like pace-makers or other healthcare equipment that do not behave according to their functional specification may cause loss of life. Even for non-critical systems, failure after shipment will result in a product recall which means wasted money and a loss of reputation for the company. The importance of functional correctness, therefore, influences system design methodology. In each step of the design, a designer needs to make sure that the model reflects the original intent of the design and that it performs efficiently, safely and successfully. This is achieved by verification of each system design model.

The techniques for verifying design models can be classified into two groups:

1. Simulation based methods

2. Formal methods

Verification techniques belonging to either of the above groups rely on the same basic principle: the implementation model must be checked to ensure that it satisfies the specification. In simulation based methods, the specification is a set of properties that the implementation model must be checked for. Some instances of these properties are expressed as pairs of stimulus and expected behavior. The stimulus forms the input to the implementation model being simulated and the expected behavior is checked by monitoring the output of the simulated model.

D.D. Gajski et al., *Embedded System Design: Modeling, Synthesis and Verification*,
DOI: 10.1007/978-1-4419-0504-8_7,
© Springer Science + Business Media, LLC 2009

In formal verification methods, a property is statically checked instead of some instances of the property. This means that once the verification process is complete, we can be assured that the implementation model satisfies the property under all inputs. There are different types of formal verification, the most popular ones being equivalence checking, model checking and theorem proving. Each of these methods expresses the specification as well as the implementation as a mathematical model.

In equivalence checking, the formulas for both the specification and the implementation are reduced to some canonical form (if one exists) by applying mathematical transformations. If their canonical forms are identical, then the specification and the implementation are said to be equivalent. In model checking, the implementation is expressed as a state transition system and the specification is a set of properties. Each property in the specification is checked by traversing all the states in the transition system. Theorem proving methods try to deduce the equivalence of formulas of the specification model and the implementation models, which are written in a given mathematical logic. Using the laws of the logic, the implementation formula can be reduced to that of the specification, or vice versa.

At first sight, simulation may seem too expensive, too time consuming or even less trustworthy than formal methods. Indeed, simulation is only a partial test since we are checking for instances of a property and not the complete property under all input scenarios. However, simulation is still the predominant technique for verification. There are various historical as well as practical reasons for this. In the first place, the application of formal methods to design verification is relatively recent compared to simulation. Hence, we have not yet seen the same scale of adoption for formal verification techniques and tools as that for simulation tools. Secondly, formal verification often forces the designer to comply with certain rules in modeling, so that the model can be easily converted to its mathematical formulation. In contrast simulation allows designers a high degree of independence in writing models. Almost any legally written code in a design language can be simulated. Thirdly, typically designers come from an engineering background and, in general, do not have the expertise in mathematical theory to efficiently use formal verification techniques.

Simulation tools' popularity and ease of use notwithstanding, the importance of formal verification in system design cannot be understated. As designs become larger and more complicated, simulation takes far too long to meet the required verification quality. As a result, there has been a push towards an efficient verification methodology to apply alongside a design methodology. Techniques like assertion based verification are being used to complement the traditional simulation and debugging of design models. Designers are employing formal methods like logic equivalence checking to minimize or even eliminate the need for costly gate-level simulations.

As seen in previous chapters, a large number of system models may be used during the design process. Verification of individual models by conventional methods alone would not be cost efficient as design moves to the system level. The sheer size of the designs prohibits exhaustive simulation. A possible direction for efficient verification is by formalize the model construction and develop develop methods to ensure correctness of model refinements. This will allow us to use conventional methods at higher levels of design abstraction, when the model complexity is still manageable.

This chapter will provide an overview of various techniques for the verification of systems, ranging from simulation based methods to formal methods. We will discuss the theory behind each technique and elucidate it with helpful examples. A comparison of the techniques is given, based on metrics like cost, applicability to the design and coverage. We then discuss the challenges in verifying large systems with traditional techniques and provide an outlook for alternatives in the future.

7.1 SIMULATION BASED METHODS

Simulation is the most widely used method to verify system models. The design to be tested is described in some modeling language and is referred to as design under test (DUT) as shown in Figure 7.1. The DUT sits in a simulation environment consisting of stimuli and monitors. The stimuli are a set of values that are applied to the DUT's inputs. These inputs then trigger a series of events and computations as described in the DUT model. It is the job of the simulator to keep track of all these events and propagate them through the DUT. This is a scenario in a typical event-driven simulator.

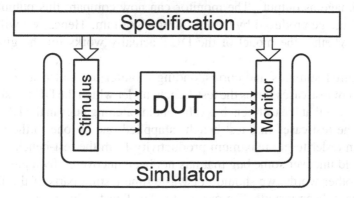

FIGURE 7.1   A typical simulation environment

As the events are propagated through the DUT, the values of various variables in the model are computed. Whenever the variables' values are updated, as a result of computation, a new event is generated to inform the simulator of this update. Consequently, the simulator executes any computation that depends on the updated variables. The output of these computations leads to newer events and so on and so forth. Eventually, the variables representing the output of the DUT are updated. This is where the monitor comes into picture.

The primary responsibility of the monitor is to make sure that the output values during simulation of the DUT match the expected expected output values. Note that during simulation the outputs may change over time. Hence, the monitor must store all expected output values along with their respective time of appearance. Once the output is updated at a given simulation time, say t, the monitor must check if this output is equal to the expected output at time t. If it is, the simulation is successful. However, if the simulated and expected values are not identical, the monitor flags an error.

TEST-BENCH

The stimuli and monitor for a verification effort are created from a high level specification of the DUT. This specification consists of properties that are expected to hold in the model. Sometimes the specification is merely a high level algorithmic description of the design. For instance, while designing a JPEG Encoder, we expect the model of the DUT to encode a bitmap image to a JPEG image. In this case, we can use the JPEG encoding algorithm as our specification. The stimulus for simulation is simply a bitmap image. The expected output can be generated by running the encoding algorithm on this bitmap image. The resulting JPEG image would thus be a reference for checking the output from simulated DUT. Once the DUT is simulated, it will produce a JPEG image as output. The monitor can now compare this output against the JPEG image produced by the encoding algorithm. Hence, we will be able to verify whether the model of the DUT actually works for the given input instance.

The paired stimulus and corresponding monitor are called a test-case. A collection of test-cases forms the test-bench, under which the DUT is simulated. It is important that the test-bench is efficient in catching bugs in the DUT model. Each of the test-cases in a test-bench is applied one by one to the simulated model. In order to get maximum productivity from the test-bench, each test-case should uncover some bug that has not been uncovered by a previous test-case. In other words, we should not waste time testing parts of the DUT that have already been tested. The part of the DUT tested by a given test-case is called its simulation *coverage*.

COVERAGE

Although a rigorous definition of simulation coverage for a test-bench is hard to come by, in general, it refers to the percentage of DUT that has been checked by the various tests applied during simulation. However, it is difficult to quantify a DUT. We can quantify it by the lines of source code for the model written in some design language. Alternately, we can use a state diagram which represents all possible scenarios that might exist during a model's execution. The DUT can, thus, be quantified by the states and transitions in the state diagram. Unfortunately, these representations are incomplete and do not truly capture the entire behavior of the design. The best bet in using coverage for generating new tests is to employ as many quantification metrics as possible.

We can use statement coverage to see how many lines of code were *visited* during a verification run. If during simulation with a given test-case, 100 statements out of 1000 statements in the design were executed, then we say that the statement coverage for the test-case is 10%. However, this is a very weak metric of coverage, since not all possible scenarios for those 100 statements were exercised. For instance, the statement

$$a = b/c$$

will execute correctly if $b = 4$ and $c = 2$, but will cause an exception if $c = 0$. Statement coverage would tell us if the given statement were executed during simulation but not advise the user to check for the corner case of $c = 0$.

In the case of state coverage, we measure the number of states and transitions that are "traversed" during the model's simulation. A state S is said to be traversed if during simulation of the DUT, S was visited at least once. This would ensure that the scenario represented by s was tested during simulation. Hence, for the above example, if we were to cover the state with $c = 0$, we would cover the overflow scenario. However, this would require that each legal value of c (and other variables) should have different states in the state diagram. Clearly, it will result in an unreasonably large state diagram.

PERFORMANCE IMPROVEMENTS

In an ideal scenario, one would like to run the minimum number of test cases to cover as much of the design as possible. However, this would require some method to estimate the coverage of test-cases and generate the test bench in an efficient manner. In the absence of this kind of dynamic coverage feedback, the author of the test-bench may choose to randomly generate test-cases. This means that the testing is not directed at finding specific bugs. Instead, the designer hopes that the random tests are fairly distributed in the range of possible inputs. Naturally, the quality of such test-cases is in general poorer compared to test-cases trying to cover particular scenarios.

The simulation performance can be improved by choosing test cases intelligently to maximize coverage with minimal simulation runs. One optimization is to reduce test generation time by giving constraints to stimuli and testing with only valid inputs. For instance, if we know from the design specification that a particular scenario is never going to occur, we do not need to spend time in writing tests for such a scenario.

Besides stimulating the design with relevant test vectors, we can improve our understanding of the design by performing white box testing. In white box testing, we also monitor the non-primary output variables in the model. Of course, such an approach only makes sense if the internal details of the design are available. Since we do not need to wait for the errors to be observed on the primary outputs, the debug time is significantly reduced because the error is usually observed close to its origin and this minimizes the effort neccessary to locate the bug.

Another strategy to improve simulation performance is to speed up the simulation itself. This is achieved by either using a faster simulation algorithm or using hardware support for testing. For cycle accurate models of the design, it is sometimes possible to use a cycle simulation algorithm over the traditional event driven simulation algorithm. Since a cycle based algorithm does not take into account every event during a model's execution, it avoids the overhead of processing each event. In the case of hardware assisted testing, a functional prototype of the design is implemented onto an FPGA. In some cases, part of the model can even be implemented on the FPGA and tested during the software simulation of the remaining design. This is achieved by using hardware emulators capable of exchanging events with the software simulator.

The speed and efficiency of simulation is critical because the rise in complexity triggers a shift to a higher level of design description. We have witnessed this in the shift from the transistor level to the gate level, RTL and now to the system level. By eliminating any unnecessary implementation details, we can describe the behavior of the intended system in a succinct and efficient model. A system level modeling language aids such functional design. Its simulators are typically several orders faster than cycle accurate simulators. Since the majority of design re-spins are due to functional errors, it is imperative that we first focus on getting the functionality of the design right, before implementing it. Hence, designs increasingly need to be modeled at higher levels of abstraction to leverage the simulation performance at the system level.

7.1.1   STIMULUS OPTIMIZATION

Another way to optimize simulation is to test only those cases that the product may actually encounter. Writing down all possible test vectors for simulation can be a painful task. Also, generating test vectors randomly might result in a lot of invalid vectors. Since the design is typically constrained to work for only select scenarios, we can use this knowledge to generate only the valid test vectors. The test scenario can thus be written in some language and a tool can be used to generate valid test vectors for that scenario. Such languages are known as verification languages. The key is to specify the property of the design to be tested along with its description. These properties specify the behavior of the design using a formal language. For example, we may specify that the value of variable y becomes 1, two clock cycles after variable x is set to 0. Then a test vector may be generated that sets x to 0 and observes the value of y after two clock cycles. Therefore, we can automatically validate the *assertion* about the behavior of the design. In short, the test generation tool analyzes the given properties and produces test vectors to validate those properties.

The constraints specified in the property lead to a set of legal inputs that form the test pattern. Some times it is not necessary to have a different language to do this because the properties can be embedded in the design model as special comments known as *pragmas*. The test generation tool can identify these pragmas and produce tests based on them. However, for the synthesis tool, the pragmas are merely comments and hence do not interfere with its operation.

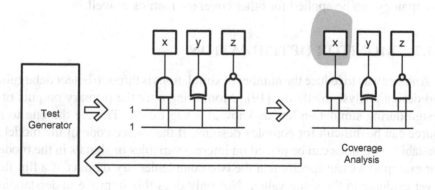

FIGURE 7.2   A test case that covers only part of the design.

Analyzing the results from coverage is another way to minimize the number of test vectors. For instance, the code coverage feedback technique can be visualized in Figure 7.2. A simulation of the model with input vector *11* results in only block x being covered. This is because block x is enabled by an AND gate whose inputs are the two signals shown in the figure. The other two blocks, y and z, are enabled by the XOR and NAND of the inputs, respectively. Thus the computation inside y and z is not triggered by this test. The coverage analysis tool thus comes back with the answer that only block x has been covered.

The designer analyzes the coverage result and comes up with a vector *10* to cover blocks y and z. The enabling inputs to y and z are set to 1, thereby

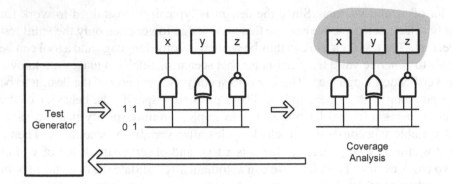

FIGURE 7.3 Coverage analysis results in a more useful test case.

enabling these blocks. Note that vector *00* would not cover block *y* and is thus not used. Therefore, the coverage feedback allows the designer to cover all blocks with just two input patterns. Without this knowledge, in the worst case one would have to use all possible input combinations to achieve complete coverage. Although this is a simple and cosmetic example, it illustrates the benefits of such a coverage feedback mechanism. Using the same principles, this strategy can be applied for other coverage metrics as well.

7.1.2 MONITOR OPTIMIZATION

Another way to reduce the number of simulations is through beteer debugging and design analysis methods [19]. Monitoring only the primary outputs of a design during simulation lets us know if a bug exists. Tracing the bug to its source can be difficult for complex designs. If the source code of the model is available, assertions can be placed on internal variables or signals in the model. For example, we can specify that the two complementary outputs of a flip-flop never evaluate to the same value. Not only does this improve understanding of the design, it also points out the bug much closer to the source. Assertions can also be used to check the validity of properties over time, such as protocol compliance. However, the designer must ensure that the assertions do not get synthesized along with the design. Therefore, they must be written either in a language different than the design, or as special comments that can be ignored by the synthesis tool.

Graphical visualization of the structure and behavior of a design also helps debugging. Specifically, visually correlating different representations, such as waveforms, net lists, state machines and code, allows the designer to easily identify design bugs and locate the source code for the errorneous part of the model. As shown in Figure 7.4, the piece of code in a model source may be

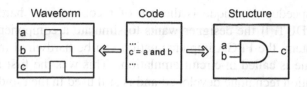

FIGURE 7.4 Graphical visualization of the design helps debugging.

visualized structurally either as a net list, shown by the AND gate, or in a waveform showing the timing behavior of the circuit. Furthermore, if these design representations are then correlated, debugging becomes significantly simpler. Therefore, designers often use graphical representations for debugging and analysis.

Different types of simulation errors are more conveniently observed in different representations. For instance, a timing error is most easily identified in a waveform display. On the other hand, a logical error can be easily identified on a gate net list. But by correlating these different representations, an error identified in a visual representation can quickly be located in the source code.

7.1.3 SPEEDUP TECHNIQUES

Overall simulation time can also be reduced by simply increasing the simulation speed. The two common speedup techniques are cycle simulation and emulation. Cycle simulation is used when we are concerned about the signal and variable values only at the clock boundaries. This improves the simulation algorithm to update signal values at clock boundaries only. In contrast, event driven simulation needs to keep track of all events, even those that between the clock edges, and is thus much slower.

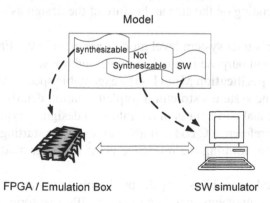

FIGURE 7.5 A typical emulation setup.

Another speedup technique is the use of reconfigurable hardware to implement the DUT. If the designer wants to simulate a component in a larger available system, the FPGA implementation can be hardwired in the system. This technique is called in-circuit emulation. This was the first hardware assisted simulation technique developed and is still used in the rapid prototyping of systems.

A different scenario for emulation is dubbed simulation acceleration. In this method, the entire system is not prototyped on the FPGA. Instead, only a part of the design is synthesized onto the FPGA board while the remaining part is still simulated in software as illustrated in Figure 7.5. One important consideration for choosing partial implementation is that the entire model of the design may not be synthesizable. Increasingly, embedded systems include a significant software component and, therefore, may not be easily prototyped on FPGA. Also, since most HDLs are not completely synthesizable most designs also contain non-synthesizable hardware. For simulation acceleration, the synthesizable part of the hardware is implemented on an FPGA while SW and the non-synthesizable HW runs on a software simulator, which talks to the emulation tool via remote procedure calls.

7.1.4   MODELING TECHNIQUES

A different approach for reducing functional verification time is to model the system at higher abstraction levels. By abstracting away unnecessary implementation details, the model not only becomes more understandable, but it also simulates faster. For instance, models with bus transactions at word level simulate faster than those at bit level because the simulator does not have to keep track of bit-toggling on bus wires. Similarly, models with coarse timing result in fewer events during simulation. There are several abstract models that we can use depending on the size and nature of the design as well as the design methodology.

Some of the abstract system level models are as follows. Each model has its own semantics and purpose in the design methodology.

Functional Specification Model is the executable specification that does not contain any of the system's structural implementation details. Its purpose is to check the functional correctness of the intended design. It typically executes at the speed of the reference C code. It also serves as the starting point for design space exploration, forming the reference point for other models in the design flow.

Platform Model considers only the partitioning of system functionality onto various system components such as processors, IPs, memories etc. The primary purpose of this model is to evaluate the HW/SW partitioning decision and to serve as an input for communication synthesis.

Transaction Level Model contains the communication structure of the intended design along with the HW/SW partitioning. The communication, however, is not yet pin accurate. Since we are only interested in the approximate timing of communication, the data transfers between components are modeled as abstract bus transactions.

These abstract system level models need not be the only ones used in a design flow. Depending on the application and design characteristics, different models may also be employed. However, the guiding principle in choosing such models is the simulation speed and the possibility for design space exploration.

7.2   FORMAL VERIFICATION METHODS

Formal verification techniques use mathematical formulations to verify designs [113]. The key difference from simulation based verification is the absence of a test pattern. The formal verification process either compares two different models of the design or shows that a property is satisfied on the model. In either case the answer from the formal verification tool is valid for all scenarios. This is one of the strongest points of formal methods; they can provide absolute answers to verification problems. On the flip side, however, most formal techniques involve converting the model to some abstract mathematical form, which may not always be feasible.

In the design industry, there are three primary types of formal verification. The first is *equivalence checking*, which can be used to compare two models. In general, equivalence checking can establish whether two models will give the same result under all possible inputs. This is particularly useful in checking the correctness of the synthesis and optimization steps. Due to the critical importance of model correctness, the designer cannot trust the synthesis tools to preserve all the properties of the original model. Hence, equivalence checking actually serves a validation of the synthesis step. For the purpose of equivalence checking, one needs to define some notion of model equivalence such as logical equivalence or state machine equivalence. Based on this notion, the equivalence checker then proves or disproves the equivalence of the original and optimized/synthesized models.

Model checking, on the other hand, takes a formal representation of both the model and a given property and checks if the property is satisfied by the model. More often than not, system models become too big or complicated at lower levels of abstraction. Hence, their behavior needs to be checked against some abstract specification. This abstract specification is essentially a set of properties that are expected to hold on the model. These properties are similar to the one described in Section 7.1. A model checking tool can automatically verify if each of these properties holds in the model. Most modern assertion based verification tools, compose these abstract properties from assertions in the model and use them for model checking. The properties are temporal in nature, i.e. they define the behavior of the system over time. A complete theory of temporal logic forms the framework on which model checking is based. These properties are written as formulas in the temporal logic and the model checker tries to prove or disprove each formula.

A somewhat different approach in formal verification, known as *theorem proving*, tries to prove properties under some mathematical logic by using deductive reasoning. Theorem proving has the advantage of being applicable to almost any domain of problems and is hence very versatile. This flexibility is also a reason for its biggest disadvantage: it is extremely hard to create automated tools for theorem proving. The basic idea behind theorem proving is to express both the specification and the implementation models as mathematical formulas. On the basis of axioms, which are established truths in the given logic, one can show the equality of the two formulas. Hence, the proof establishes that the implementation model is a valid substitution for the specification.

In this section we will look at all these formal verification methods in detail.

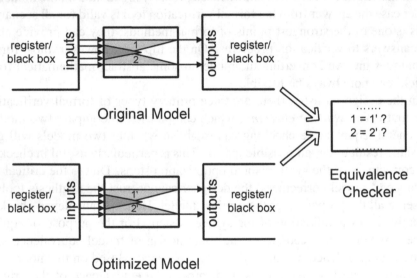

FIGURE 7.6 Logic equivalence checking by matching of cones.

7.2.1   LOGIC EQUIVALENCE CHECKING

During the optimization of logic circuits, the design is optimized to reduce the number of gates which thereby reduces circuit delay. The designer is responsible for the logical correctness of any such transformation. A logic equivalence checker verifies that the result of the synthesis or optimization is equivalent to the original design. This is achieved by dividing the model into logic cones between registers, latches or black-boxes as shown in Figure 7.6. The combinational part between registers in an RTL or Gate model has as many logic cones as the number of its outputs. After synthesis, as the combinational part is optimized, the logic cones change their structure, but it is still possible to correlate these cones in the original model and those in the optimized one.

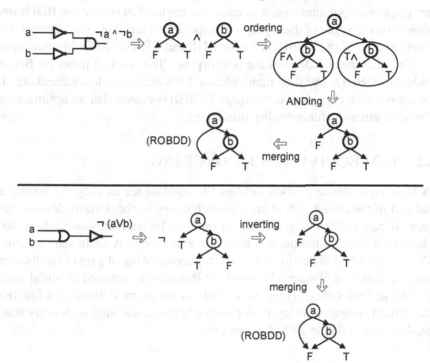

FIGURE 7.7 DeMorgan's law illustrated by ROBDD equivalence.

These logic cones are nothing but combinational circuits and can, thus, be described with Boolean expressions. Since the registers stay the same, we are only interested in knowing whether the optimization on the combinational circuit is correct. Therefore, we need to compare the Boolean formulas for the corresponding logic cones in the two models. This is made possible by creating directed acyclic graph representations of Boolean functions. These graphs, known as Reduced Ordered Binary Decision Diagrams (ROBDD), have a special property in that they define a canonical form for a given Boolean function [27].

Moreover, if two Boolean functions are equivalence, they will have isomorphic ROBDDs. Consequently, we can reduce the original and optimized cones to their respective canonical forms and check if they are isomorphic. Figure 7.7

illustrates this principle on the DeMorgan's law for Boolean functions, which states

$$!(a + b) \;=\; (!a).(!b)$$

So ROBDDs are a compact way of representing Boolean functions. Furthermore, all Boolean functions, such as conjunction (AND), disjunction (OR), and negation (NOT) may be expressed as graph manipulations of ROBDDs. Other graph manipulations, such as merging, are used to *reduce* the BDDs into canonical form. Some of these graph manipulations are shown in Figure 7.7 in the construction of the ROBDDs for the LHS and RHS of the DeMorgan equation. Note that these ROBDDs are isomorphic. The seminal paper by Bryant introduces ROBDDs and their manipulation for logic equivalence checking. In logic equivalence checking, isomorphic ROBDDs ensure that an optimization of the logic circuit is functionality preserving.

7.2.2   FSM EQUIVALENCE CHECKING

A logic equivalence checker verifies the equivalence of only the combinational part of the circuit. There are also techniques to check equivalence of the sequential part of the design [135]. In order to understand those techniques, we have to define the notion of a finite state machine. A finite state machine (FSM), as described in Section 3.1.2, is a tuple consisting of a set of inputs, a set of outputs, and a set of states. Some of the states are designated as initial states and some as final states. Transitions between states are defined as a function of the current state and the input. An output is also associated with every state. Formally, we can define a FSM as the tuple

$$< I, O, Q, Q0, F, H >, where$$

I is the set of inputs O is the set of outputs Q is the set of states $Q0$ is the set of initial states F is the state transition function Q X I *if* Q H is the output function Q *if* O

We may think of a FSM as a language acceptor. We further define Qf as the set of final states. If we start from an initial state (in $Q0$), supply input symbols from a string S and reach a final state, then S is said to be accepted by the FSM. The set of all acceptable strings forms the language of the FSM.

We can also define the notion of a FSM product. The product of two FSMs $M1$ and $M2$ has the same behavior as if $M1$ and $M2$ were running in parallel. Therefore, given FSMs $M1$ and $M2$, such that

$$M1 \;:\; < I, O1, Q1, Q01, F1, H1 >,$$
$$M2 \;:\; < I, O2, Q2, Q02, F2, H2 >$$

The product FSM $M1 * M2$ may be written as

$$M1 * M2 \;:\; < I, O1 U O2, Q1 X Q2, QO1 X QO2, F1 X F2, H1 X H2 >\;.$$

The total number of states in the resulting machine is the product of the number of states in each machine. The product machine carries all possible pairs of states, one from each of the two input machines. The paired states are labeled with the pair of corresponding outputs as well. The inputs on the transition arcs are also pairs of possible inputs from each machine.

Using the above definitions, we can define sequential equivalence of FSM models through a simple metric. We must prove that if two FSMs are given the same inputs in the same sequence, then under no circumstances would they produce different outputs [48]. Only then can we claim that the machines are equivalent. The specification and its implementation are both represented as FSMs Ms and Mi respectively. We must ensure that the input and output alphabet of the two machines is the same.

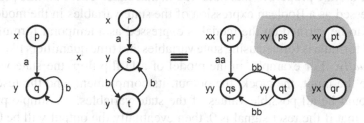

FIGURE 7.8 Equivalence checking of sequential design using product FSMs.

To perform FSM equivalence checking, we first derive the product machine $Ms X Mi$. Now all the states in $Ms X Mi$ that have a pair of differing outputs are labeled as final states as shown in Figure 7.8. Ms has two states, p and q, while Mi has three states, r, s, and t. States p and r produce output x while the other three states generate output y. Therefore, in the FSM product, the states ps, pt and qr have output pairs with different symbols (xy or yx) and are thus labeled as final states. We also keep only those transitions that have the same symbols in the input pair. What we are trying to prove is that for the same sequence of inputs, Ms and Mi would produce the same sequence of outputs. In other words, we should never reach a state with a pair of non-identical outputs. Since such states are the final states in the product FSM, they should never be reached if Ms and Mi are equivalent. Therefore the product FSM should not accept any language.

The case of non-equivalent FSMs is shown in Figure 7.9. Here, we can see that the state qt in the product FSM produces different outputs (y and x) for the two FSMs amongst the final states in the product FSM reachable from the start state pr. Therefore the two FSMs can possibly produce different outputs for the same stream of inputs and are not equivalent.

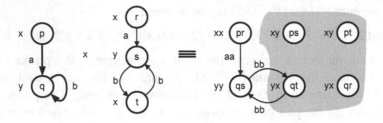

FIGURE 7.9 Product FSM for with a reachable error state.

7.2.3 MODEL CHECKING

Model checking [42] is another formal technique for property verification. In it, a model is represented as a state transition system, which consists of a finite set of states, transitions between states, and labels on each state. The state labels are atomic properties that hold true in that state. These atomic properties are expressed as a Boolean expression of the state variables in the model. The property to be verified on the model is expressed as a temporal formula. The temporal formula is formed using state variables and time quantifiers like *always* or *eventually*. For example, in the model of a D-flip flop, the state variables would be the input, the clock, the output, its complement, and the reset. The states would be all possible values of the state variables. A simple property might be that if the reset signal is 0, then eventually the output will be 0.

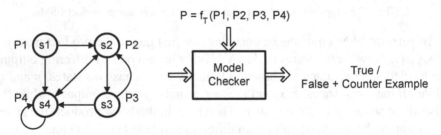

FIGURE 7.10 A typical model checking scenario.

Figure 7.10 shows a typical model checking scenario. The model checker works on the state transition system of the model and the given property and produces a result *TRUE* if the property holds in the model. If the property does not hold, the checker gives a counter-example to show that the property is violated. This feature of model checking is very helpful in debugging because it provides a readymade test case. In the given figure, we see the state transition system of model *M* and a temporal property composed from the properties of the individual states.

The idea behind model checking can be visualized by unrolling the transition system. We start with the initial state and form an infinite tree, called the computation tree, as shown in Figure 7.11. In Figure 7.10, we can see the transition system of the design which is being input to the model checker. This transition system has its start state as *S1*. Therefore, the computation tree for this transition diagram is rooted in *S1*. Starting from *S1*, we traverse the outgoing arcs to reach other nodes of the transition system. This breadth first traversal of the transition systems leads to the computation tree.

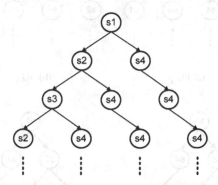

FIGURE 7.11 A computation tree derived from a state transition diagram.

The state traversal of the transition diagram represents the behavior of the model as time progresses. If one were to start from the root of the computation tree and follow some path down the tree, one would actually be executing some possible behavior of the design. This notion allows us to define useful temporal properties using the computation tree. Temporal properties are properties that hold for some given time as defined by the temporal operators.

Intuitively, we can consider properties that will hold all the time or some time in the future. These temporal notions are written using letters *G* and *F* respectively. For example, *Gp* means that property *p* is always true. Similarly, *Fp* means that property *p* will eventually hold true, sometime in the future. Other factors for temporal properties may be whether the properties hold on all paths of computation or only one path. These factors are represented by letters *A* and *E* respectively. Using these temporal operators and the computation tree, we can define a myriad of temporal properties, as shown in Figure 7.11. The various temporal formulas are illustrated on the computation tree. The states in which the property holds true are represented by shaded nodes.

In Figure 7.12(a), the property *p* holds true all the time on one path. This is realized by combining the E and G operators to get the formula *EGp*. Similarly, *AGp* in Figure 7.12(b) shows that p is true all the time on all the paths in the model. By the same principle, we can derive partial computation tree for *EFp* and *AFp*, as shown in Figure 7.11(c) and (d).

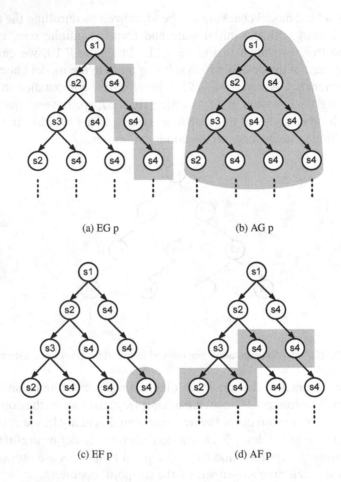

(a) EG p   (b) AG p

(c) EF p   (d) AF p

FIGURE 7.12 Various temporal properties shown on the computation tree.

Although automatic model checking provides the advantage of comprehensive property veriification, it suffers from the serious problem of state space explosion. The state transition system grows exponentially with the number of state variables. Therefore, memory for storing the state transition system becomes insufficient as the design size grows. Atypical modern RTL design has hundreds of state variables, at the very least. This means that the number of possible states in the transition diagram would be at least 2^{100}. This many nodes (and their related data structure) would be impossible to hold in memory. This is one of the reasons why model checking is very effective for control oriented designs, but performs poorly on data intensive designs. However, there have been significant research efforts to alleviate the state explosion problem, as we shall discuss later in this section.

7.2.4 THEOREM PROVING

An alternative approach to formal verification is verification by deductive reasoning. Using this technique, the specification and implementation models are written as formulas in some mathematical logic. Then a theorem is established and proven for the equivalence of these formulas. If a proof is found, the models are equivalent. However, if a proof is not found then the equivalence of models is inconclusive [78].

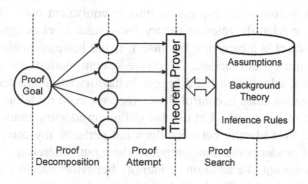

FIGURE 7.13 Proof generation process using a theorem prover.

The proof uses certain assumptions about the problem domain and axioms of the mathematical logic. In the domain of circuit design, an assumption might be that the power supply is always at logic level 1 while the ground is logic 0. The proof is constructed by breaking down a complex proof goal into smaller goals as shown in Figure 7.13. The smaller goals are then simplified using assumptions and then passed onto an automatic theorem prover.

Theorem proving is still a largely manual process. Several steps of simplifying and breaking down proof goals may be required before an automatic prover can solve it. Typically the original theorem checking a property is a very complex formula. The formula is decomposed into smaller formulae and then an attempt is made to automatically check correctness of the decomposed formulae. In the worst case, this decomposition process must be repeated several times before the formulae become simple enough to be proved by the automatic tool.

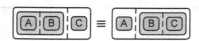

FIGURE 7.14 Associativity of parallel behavior composition.

Here, we present a simple example of the use of theorem proving for verifying the associativity of a parallel composition, as shown in Figure 7.14 in which parallel behaviors *A*, *B*, *C* are combined in two different fashions. On the left side, behaviors *A* and *B* are combined into a parallel behavior which is then combined into another parallel behavior with *C*. On the right side, behaviors *B* and *C* are combined first and the resulting behavior is then combined with *A*. The example demonstrates the basic principle of associativity as applied to the parallel composition of behaviors.

The proof must take place under a given theory. A theory involves objects and composition rules that are used to create expressions. Also, there must be basic laws to convert an expression into an equivalent one. Therefore, we must determine what the relevant theory for system level design is. Clearly, the primary object is a behavior, because it is the basic computation element required to build system models. To create system models, we need composition rules that allow us to build bigger behaviors from smaller ones. One of the composition rules is control flow, since we need to define an order for the execution of behaviors. Let us also define an identity element in the set of behaviors. This identity behavior does not perform any computation and, hence, always produces the same output as the input. Therefore, from a purely functional viewpoint, the addition of identity behaviors would not modify the execution of the model. The basic laws for our theory of system models are shown in Figure 7.15.

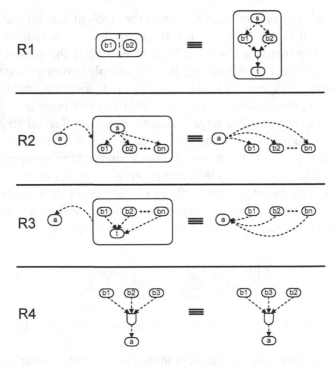

FIGURE 7.15 Basic laws for a theory of system models.

With these laws in place, the proof process takes the implementation formula and reduces it to the specification formula by a number of proof steps. Each proof step uses an assumption, an axiom, or an already proven theorem. In our case, the axioms are the laws as shown in Figure 7.15. Our proof goal is to show the equivalence of parallel compositions under associativity. If the function $par(b1; b2 ::: bn)$ represents a parallel composition of n behaviors, we can write our proof goal as the following equation:

$$par(par(a; b); c) \quad = \quad par(a; par(b; c))$$

The proof steps use the basic laws of our theory, as presented earlier, to transform the expression on the RHS into the expression on the LHS.

7.2.5   DRAWBACKS OF FORMAL VERIFICATION

Compared to simulation based methods, formal verification methods have not been as well accepted in the industry due to several drawbacks. Logical equivalence checking works only for combinational logic while FSM equivalence checking requires both the pecification and implementation machines to have the same set of inputs and outputs.

Model checking, besides suffering from the state explosion problem, is not suitable for all types of designs. Since it needs a state transition system, it works best for control intensive designs such as bus controllers. Automatic theorem proving has not become very popular in the industry either; the foremost reason for this is the amount of manual intervention required in running the theorem proving. Since different applications have different kinds of assumptions and proof strategies, it is infeasible for a theorem proving tool to generate the entire proof automatically. Secondly, most designers lack a background in mathematical logic. Therefore, it requires a huge investment and long training time for them to start using theorem proving efficiently.

7.2.6   IMPROVEMENTS TO FORMAL VERIFICATION METHODS

Recently, tools vendors and academics have made several improvements to formal techniques, particularly in model checking. Symbolic model checking [136] encodes the state transition system using BDDs, which is a much more compact representation than exhaustively enumerating the states and transitions. Since BDDs represent sets of states, the model checking algorithm can operate on sets of states rather than individual states.

Another innovation is bounded model checking, which checks if a model satisfies a property on paths of length at most K. The number K is incremented until a bug is found or the problem becomes intractable. Partial order reduction techniques are usually used in model checking for asynchronous systems, in which concurrent tasks are interleaved rather than executed simultaneously. It uses the commutativity of concurrently executed transitions, which result in the same state when executed in different orders.

Abstraction technique is used to create smaller state transition graphs. The specified property is described using some state variables. The variables that do not influence the specified property are eliminated from the model, thereby preserving the property while reducing the model size.

7.2.7   SEMI-FORMAL METHODS: SYMBOLIC SIMULATION

Semi-formal verification refers to the ues of formalisms and formal verification methods in a simulation environment. The idea behind symbolic simulation is to significantly minimize the number of simulation test vectors by using symbols to achieve the same coverage. In symbolic simulation, the stimulus applies Boolean variables as inputs to the simulation model. This is illustrated in Figure 7.16. During simulation, the internal variables and outputs are computed as Boolean expressions. In order to check for correctness, the output expression is compared with the expected output expression as defined by the Monitor. BDDs can be used to store the Boolean expressions in the Monitor. The BDDs of equivalent Boolean expressions can be reduced to identical canonical forms. Therefore, the equivalence of a specified output expression to a simulated output expression can be checked easily. For larger circuits, in which the BDD size may blow up, we can use SAT solvers as is increasingly common.

7.3   COMPARATIVE ANALYSIS OF VERIFICATION METHODS

Different application domains and types of systems may require different verification methods. Formal methods, though time consuming and difficult to deploy, may be needed for ASIC implementation of mission-critical systems or processors because of the thoroughness of the verification they perform. On the other hand, inexpensive reconfigurable devices may not require such exhaustiveness, so randomized simulation may be sufficient. It is important to consider how to best introduce verification in a system design flow. Depending on the abstraction level of the models and the application characteristics, different verification techniques may be employed.

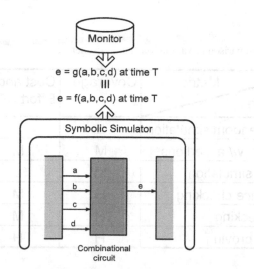

FIGURE 7.16 Symbolic simulation of Boolean circuits.

In order to determine the most suitable verification method, one can define some metrics to evaluate them. The three most common metrics that we discuss here are coverage, cost, and scalability. Coverage of a verification method determines how much of the design functionality has been tested. Cost includes the money spent on the purchase of tools, hiring of experts, and training users. Scalability refers to any limitations on the size or type of design that we are verifying.

Formal verification claims to provide complete coverage. However, the coverage is limited to the given property and the model representation. For instance, model checking covers all possible states in the state transition representation of the model for a given property. Logic equivalence checking covers the combinational part of the model only. Nevertheless, the coverage of formal methods, if they are applicable, is significantly greater than that of simulation methods over the same run-time.

Using assertions in the design can help make better test cases because exercising the assertions ensures that the tests are useful and valid. Pseudo-random testing, on the other hand, would wastefully generate test inputs that are invalid for the design.

The cost and effort involved in a verification method also influences the design phase in which the method is used. For instance, the preliminary phase usually employs random simulation to uncover most of the egregious bugs because most designers have experience with simulation tools and debuggers making it cost effective at this stage. Designers might also employ assertions to generate more directed tests and to verify correctness of known corner cases. As

TABLE 7.1 A comparison of various verification schemes.

Metric Technique	Coverage	Cost and Effort	Scalability
Pseudo random simulation	L	L	H
Simulation w/ assertions	M	M	H
Symbolic simulation	M	L	L
Equivalence checking	H	M	M
Model checking	H	M	L
Theorem proving	H	H	M

the verification process continues, however, and bugs become harder to find, more expensive, specialized techniques such as model checking or theorem proving may be neccessary.

The performance of a verification method on different sizes and types of models determines its scalability. A comparative analysis of various verification schemes, based on our metric, is shown in Table 7.1. Some methods like logic equivalence checking may be limited to RTL models or below. Similarly, model checking is constrained by the number of state variables in the model. Compared to other techniques, simulation scales very well; almost any executable model at any level of abstraction can be simulated.

If we look at the trend in the acceptance of verification techniques in the industry, we find that methods with a severe drawback have been generally avoided. Model checking suffers from poor scalability and theorem proving is much too expensive, thereby making equivalence checking the most commonly used technique in the industry. Likewise, assertion based techniques may require extra cost but they are replacing pseudo random simulation because they offer better coverage. A number of new verification and assertion languages are testimony to this fact.

7.4   SYSTEM LEVEL VERIFICATION

The formal verification methods discussed so far are applicable to traditional system models at the cycle accurate level or below. As the design abstraction level rises, system level models are being used increasingly for validation. During system level design, these models are refined into cycle accurate models as discussed in Chapters 4, 5 and 6. As a result, we are faced with the problem

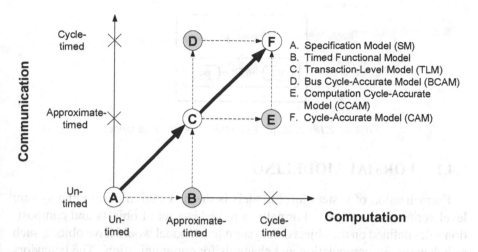

FIGURE 7.17 System level models.

of verifying the equivalence of system level models and cycle accurate models. Existing formal verification methods such as logic/FSM equivalence checking and model checking cannot be applied because system level models have not been defined formally. In this section, we will provide a brief overview of existing system level models and discuss new directions in formal system level verification.

A system level design methodology starts with a well defined executable specification model that serves as the golden reference. The specification is gradually refined to a cycle accurate model that can be fed to traditional simulation and synthesis tools. The gradual refinement produces some intermediate models depending on the choice of methodology. The details that are added to models during refinement depend on the design decisions. Each decision corresponds to one or more model transformations. If all the transformations are formally defined, the refinement process can be automated.

In general, system level models can be distinguished by the timing accuracy of communication and computation. In the graph, shown in Figure 7.17, the two axes represent the granularity of communication and computation. The functional specification at the origin is untimed, with only a causal ordering between tasks. On the other end is the cycle accurate model. A system level methodology takes the untimed specification to its cycle accurate implementation. The path through the intermediate models determines the refinements that need to be performed.

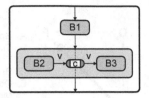

FIGURE 7.18   A simple hierarchical specification model.

7.4.1   FORMAL MODELING

Formalization of system level models is the first step in developing system level verification methods. Formally, a model is a set of objects and composition rules defined on the objects. A system level model would have objects such as behaviors for computation and channels for communication. The behaviors can be composed as per their ordering. The composition creates hierarchical behaviors that can be further composed. Interfaces between behaviors and channels or amongst behaviors themselves can be visualized as relations. A simple model using objects and composition rules is shown in Figure 7.18. The model is specified as a hierarchy of behaviors, in which hierarchy is expressed by encapsulating behaviors inside larger boxes. The various arcs show compositions amongst behaviors and channels. For example, behaviors *B2* and *B3* run concurrently, with channel *C* passing a message with variable *v* from *B2* to *B3*. Parallel composition of *B2* and *B3* executes after *B1* is finished. These compositions lead to control and data flow in the model.

A transformation of a model can be expressed by rearranging and replacing objects. For instance, in order to distribute the behaviors in a specification onto components of the system architecture, we need to rearrange the behaviors into groups. In order to use IP components, we need to replace behaviors in the model with an IP from the library. Each of these transformations has to be proven correct using a formal notion of equivalence.

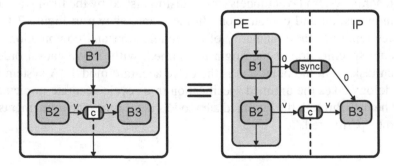

FIGURE 7.19   Behavior partitioning and the equivalence of models.

Intuitively, we can draw an analogy between the distributive law for natural numbers and the 'distribution of behaviors on different components. The distribution of multiplication over addition can be written as:

$$a * (b + c) \;\; = \;\; a * b + a * c$$

This forms a basic axiom in the theory of natural numbers. Just as the expression on the LHS is equal to that on the RHS in the distributive law equation, we can demonstrate that a model on the LHS is equal to a model on the RHS in Figure 7.19. The model on the LHS has a sequential composition of a leaf level behavior B1 and a hierarchical concurrent composition of *B* and *B3*. Channel *c* is used to send data from *B2* to *B3*. On the RHS, the model is transformed to create a concurrent composition at the top level by isolating *B3* into the independent behavior *IP*. However, the syntactic transformation does not change the function of the model. The equality is determined by the order in which the behaviors execute.

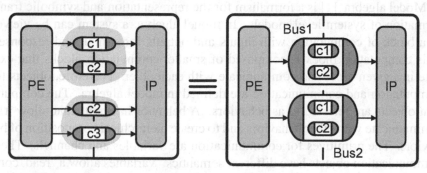

FIGURE 7.20 Equivalence of models resulting from channel mapping.

Another designer decision would be to map the abstract data channels to system buses in order to implement the inter-component communication. To reflect these decisions, we need to perform certain model transformations. These transformations would include the grouping of abstract channels as per the bus mapping and creation of hierarchical channels as shown in Figure 7.20. The hierarchical channels represent the system level bus architecture. Eventually, these hierarchical channels need to be replaced with bus protocol channels and drivers need to be added in components to implement the data transfer. The grouping transformation can be seen as analogous to the associative rule for addition of natural numbers which can be written as:

$$a + b + c + d \;\; = \;\; (a + b) + (c + d)$$

Irrespective of how we group the summation terms, the result would always be the sum of all the numbers. Similarly, irrespective of how the abstract channels are grouped in the transformed model, they would perform the same data transactions as in the original model.

The system level verification problem is to determine if the model refinements used to create a cycle accurate model from a system level model are functionality preserving. This can be achieved by creating by formalizing system level models and defining functionality preserving transformation rules. Next, each model refinement can be expressed as a well defined sequence of transformations. If each transformation is proved to be functionality preserving, the refinement will produces an output model that is equivalent to the input model. Using this technique we can solve the system level verification problem.

Let us now look at a formalism for system level models, called *Model Algebra*, that enables system level verification.

7.4.2 MODEL ALGEBRA

Model algebra [1] is a formalism for the representation and symbolic transformation of system level models. In model algebra, a system can be viewed as a block of computation with inputs and outputs for stimuli and responses. This computation block is composed of smaller computation blocks that execute in a given order and communicate with each other. Therefore, objects for computation and communication are defined in model algebra. The computation objects are referred to as behaviors. A behavior has ports that allow it to communicate with other behaviors and to create hierarchical composition of behaviors. The primitives for communication are variables and channels. These communication objects have different semantics. Variables allow a "read, compute and store" style of communication, while channels support a synchronized handshake style of communication. Composition rules are used to create an execution order for behaviors and to bind their ports to either variables or channels. In model algebra, a system is thus represented as a hierarchical behavior composed of sub-behaviors communicating via variables and channels. The objects of model algebra can be defined using the tuple

$$< B, C, V, I, P, A >$$

in which B is the set of behaviors, C is the set of channels, V is the set of variables, I is the behavior interface, P is the set of behavior ports, and A is the set of address labels for links that go over channels. We also define a subset of B representing the set of identity behaviors. Identity behaviors are those behaviors that, upon execution, produce an output which is identical to their input. We further define Q to be the subset of V such that all variables in Q are of type Boolean.

A control flow relation (Rc) determines the execution order of behaviors during model simulation. We write the relation as

$$q : b1\&b2\&...\&bn > b$$

in which b, $b1$ through bn are in B, q is in Q. The relation implies that b executes after all the behaviors $b1$ through bn, have completed and condition q evaluates to TRUE. The variable read is expressed as $v \rightarrow b < p >$, implying that behavior b reads variable v via port p. Similarly, a variable write can be expressed as $b < p > \rightarrow v$. Variable read and writes are non-blocking. Synchronized channel transactions can be written as

$$c < a >: b < p > | \rightarrow b1 < p1 > \quad b2 < p2 > ...\&bn < pn >$$

in which $b < p >$ is the out-port of the sending behavior and $b1 < p1 >$ through $bn < pn >$ are the in-ports of the receiving behaviors. The transaction takes place over channel c and uses the link address a. Channel read/writes are blocking. Both variable and channel reads/writes have corresponding relations for port mapping to create hierarchical executable models. Therefore, if a port p of behavior b is used to write to a variable, then a sub-behavior of b may only write to p using a non-blocking write relation. Finally, a grouping relation of behaviors, variables, channels and their relations are used to create a hierarchical behavior. For example, behavior b can be written as a sequential composition of $b1$ and $b2$ as follows

$$b = [b1].[b2].1 : b1 > b2$$

Transformation rules in model algebra are used to create hierarchy, flatten behaviors, resolve channel transactions into variable read/writes and control dependencies, optimize or introduce identity behaviors, and add or optimize control dependencies. Building on these transformation rules, we can apply and verify useful model refinements such as partitioning, scheduling, and routing.

7.4.3 VERIFICATION BY CORRECT REFINEMENT

In a model refinement based system level design methodology, each model produced by a refinement is equivalent to the input model. As shown in the Figure 7.21, designer decisions are used to add details to a model to refine it to the next lower level of abstraction. Each designer decision corresponds to several transformations in the model. The transformations would either rearrange the computation and communication objects or replace an object in the model with one from the library.

The notion of model equivalence comes from the simulation semantics of the model. Two models are equivalent if they have the same simulation results.

This translates to the same (or equivalent) objects in both models and the same partial order of execution between them. Correct refinement, however, does not mean that the output model is bug free. We also need to use traditional verification techniques on the specification model and prove the equivalence of objects that can be replaced by one another.

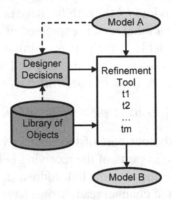

FIGURE 7.21 Model refinement using functionality preserving transformations.

Since models can be expressed as formulas, they can be manipulated according to the proven transformations of model algebra. These manipulations would allow us to have equivalent models at different levels of abstraction. Hence the debugging and verification effort can be spent only on the simplest and most abstract specification model. All subsequent models that are refined from the specification model can be proved equivalent to the specification using the rules of model algebra. Since we do not need to simulate all models exhaustively, the verification time is greatly reduced. Figure 7.21 illustrates such an approach in which detailed models are refined from abstract models using a sequence of functionality preserving transformations.

Verification may also interact with refinement in such a design methodology. This type of verification tool may be used to abstract the input and output models into model algebraic expressions. Such an abstraction would be possible in the verification semantics of the system level models are well defined. Once the model algebraic expressions are obtained, the a sequence of transformations may be used to reduce input model *A* to the expected output model *B*. If the model algebraic representation of the expected model is identical to the refined model, then the refinement is functionality preserving. Hence, model algebra enables a practical system level verification methodology that will improve designer productivity, reduce bugs and lead to more reliable embedded systems in the future.

7.5 SUMMARY

We have looked at several verification techniques ranging from simulation based methods to formal verification techniques. We also offered a comparative analysis of the various techniques and projected the future trend for system level verification. As the size and complexity of designs increase, traditional techniques might not be able to keep pace. A system design methodology with well defined model semantics may be a possible solution to the problem.

New challenges to the verification of embedded systems result from the growth in size and complexity of designs. Individually verified components do not work together due to interface issues. Also the sheer size of designs makes cycle accurate modeling and exhaustive simulation too expensive and time consuming. To answer this challenge, we must develop a comprehensive and formal system level design methodology which will require formal semantics for system level models. Furthermore, we must define methods for functionality preserving refinement of models from one abstraction level to the next. As a result, traditional simulation based verification methods can still be used for system specification model while correct refinements will avoid the need to simulate lower level cycle accurate models.

Specifying the design at a higher level of abstraction would also make traditional simulation and debugging feasible because of the smaller model size. Well defined model semantics would make it possible to define and prove correct transformations for automatic model refinement. Therefore, model formalization would make complete system verification much faster.

Chapter 8
EMBEDDED DESIGN PRACTICE

Both commercial and academic tools are available for the design of embedded systems. These tools come in three categories: system-level design, software design, and hardware design.

In this chapter, we will discuss the tools and frameworks available for these various examples of system design. We will also present examples of embedded system design and results for applications, such as JPEG encoder and an MP3 decoder. These results demonstrate the potential impact of the embedded system modeling, synthesis and verification technologies that have been discussed in this book.

8.1 SYSTEM LEVEL DESIGN TOOLS

The semiconductor revolution would not have been sustainable without the help of Electronic Design Automation (EDA) tools. Historically, the break-through of EDA came with the availability of the first Computer-Aided Design (CAD) tools for hardware synthesis (see Section 8.3). As we move to higher and higher levels of abstraction, new classes of tools gradually emerged with each new level. In recent years, we have seen a push towards development of so-called Electronic System-Level (ESL) tools. However, while there are many approaches that claim to provide ESL solutions, such as C-to-RTL tools implementing high-level synthesis of a single hardware unit (described in more detail in Section 8.3), true system-level solutions have to span the complete design space across hardware and software boundaries.

As described in detail throughout this book, a system-level design flow is typically separated into two parts: a frontend and a backend. The system design

D.D. Gajski et al., *Embedded System Design: Modeling, Synthesis and Verification*,
DOI: 10.1007/978-1-4419-0504-8_8,
© Springer Science + Business Media, LLC 2009

frontend takes a description of the application and target architecture at its input. Applications are given in some MoC to describe the desired system behavior to be implemented. Target architectures can be given in the form of architectural constraints, associated parameters, architecture templates or complete pre-defined system-level netlists. In the frontend, application computation and communication is then mapped onto and implemented on the selected or synthesized target architecture. In the process, Design Space Exploration (DSE) is performed to optimize design metrics under a set of constraints. At the output of the frontend, models of the system at various levels of abstraction are generated for virtual prototyping of the system design. Predominantly, such system models will be TLMs described in some SLDL such as SystemC. Models can be simulated or analyzed to provide feedback about the feasibility and quality of the generated design. In addition, modeling guidelines such as the SystemC TLM standard [150] promise to enable easy exchange of component or design models between companies or design divisions and across tool chains.

In the backend, high-level system descriptions are then further synthesized down to a hardware or software implementation for each PE in the system. ESL design flows thereby rely on the availability of corresponding software or hardware synthesis tools (see Section 8.2 and Section 8.3, respectively). On the software side, final target binaries for each processor are produced. On the hardware side, high-level synthesis of behavioral, C-based component models down to RTL descriptions is performed. In both cases, synthesized PE models can be re-integrated into system TLMs for cycle-accurate co-simulation with the rest of the systems. On the software side, binaries are executed in an ISS that is integrated into the overall system simulation environment. On the hardware side, RTL or gate-level models in SLDL form are inserted for this purpose. As a result, a virtual prototype of the system platform is generated.

In the end, however, the desired result at the output of a system-level design flow is a physical system prototype or a system implementation that is ready for further manufacturing. Therefore, generated software binaries should be ready to be directly loaded into target processors and RTL models should be created in the form of standard HDL code (e.g., VHDL or Verilog) such that they can feed into traditional logic and physical synthesis processes.

Overall, being based on existing commercial or proprietary backend tools, the goal of system-level design tools is to develop and apply design automation techniques to the steps in the frontend. At any level, the first set of tools to always emerge are modeling and simulation solutions that allow designers to capture models and execute them in a validation environment. Consequently, most currently available commercial system-level approaches are focused on providing models and simulators either at the application, SLDL/TLM or HDL/RTL/ISS level. Looking ahead, academic research, in contrast, is aimed at the development of subsequent system-level synthesis and verification tools, which build on modeling solutions to provide an automated path from abstract system specification down to synthesized system models and eventually a system prototype or implementation.

8.1.1 ACADEMIC TOOLS

METROPOLIS

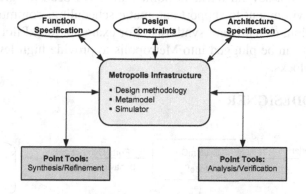

FIGURE 8.1 Metropolis framework

Metropolis [12] is a modeling and simulation environment originally developed at UC Berkeley. Metropolis is based on a Platform-Based Design (PBD) paradigm (Figure 8.1) [164] in which the target system architecture, called a platform, is assumed to be given or at least significantly pre-determined at the input of the system design flow. This constrains and simplifies the design space exploration process. In addition, a pre-defined and pre-determined platform facilitates the reuse of common design patterns across different design instances. Therefore, PDB follows a meet-in-the-middle approach and the system design problem is reduced to the mapping of a desired function onto the given target platform to create a specific design instance.

Metropolis provides a general, proprietary metamodeling language that is used to capture separate models for functionality (system application behavior), architecture and their mapping. The metamodel employs a fundamental discrete event-based execution model with concurrent processes communicating through channels (called media). In a similar manner to other SLDLs, functionality is described in the form of event-driven process networks that are general in the sense that many classes of MoCs can be represented. In addition, functionality can be annotated with non-functional constraints. The architecture is defined by using processes and media to describe available services (e.g., tasks) and resources (e.g., CPUs, memories or buses), respectively. Quantities can be associated with the architecture to model metrics such as delays. Finally, given a system functionality and architecture, synthesis or refinement is performed by defining a mapping between the two in the Metropolis metamodel as a set of additional constraints synchronizing their event execution.

Metropolis itself does not define any specific design tools but rather a general framework and language for modeling with support for simulation, validation and analysis of models. Metropolis includes a frontend for parsing of metamodels and a backend for translation of metamodels into C++/SystemC simulation code. In addition, several backend point tools have been integrated into the Metropolis environment to support automatic scheduling, communication design, verification, or hardware synthesis. For example, the xPilot system (see Section 8.3.1) can be plugged into Metropolis to provide high-level synthesis of hardware blocks.

SYSTEMCODESIGNER

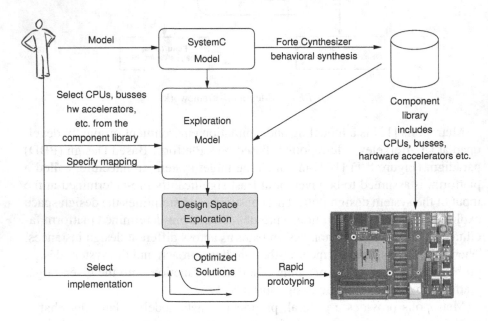

FIGURE 8.2   SystemCoDesigner tool flow

SystemCoDesigner is a system-level design space exploration environment developed at the University of Erlangen-Nuremberg in Germany (Figure 8.2) [105]. At its input, SystemCoDesigner supports applications written in a dynamic dataflow oriented MoC targeted towards streaming applications. Such input models are captured using a well-defined subset of SystemC called SysteMoC. In SysteMoC, applications are modeled as a graph of atomic actors that communicate via FIFO queues. Internally, the behavior of each actor is described in the form of an FSMD. In contrast to SDF models, SysteMoC supports applications in which actor production and consumption rates can vary dynamically at runtime. Thus, the SysteMoC model is similar to a KPN with the restriction of atomic process executions.

Once the application has been defined, SystemCoDesigner will automatically generate a library of software and hardware implementations of all actors. Software implementations are created through simple transformation of the SysteMoC input into C code. On the hardware side, Forte's Cynthesizer tool (see Section 8.3.2) is used for high-level synthesis of all actors down to RTL descriptions. All generated actor implementations are stored in a component library and are annotated with performance, area and other metrics obtained during synthesis.

Given an application, the annotated component library, and an architecture template, SystemCoDesigner can perform a fully automatic, multi-objective exploration of the design space. With the architecture template, the designer can thereby constrain the search space and restrict possible target architectures in terms of the number and type of available processors or the allowed mappings of actors to processor types. Design space exploration is performed using genetic algorithms to drive and guide the automatic search process. For every new candidate architecture selected by the search, a SystemC performance TLM is automatically generated and simulated. The generated virtual architecture model represents the mapping and scheduling of actors on the selected processors, where actors are annotated with corresponding estimated performance metrics from the component library. Simulation results are then fed back into the search algorithm to evaluate the current design point and direct the next iteration of the evolutionary exploration process.

As a result of the exploration process, a set of Pareto-optimal design solutions is obtained and presented to the user. From this optimal set, the designer can visualize the design space and subsequently select an applicable implementation option. After an architecture has been chosen, SystemCoDesigner can prototype the selected implementation on a Xilinx FPGA platform. The platform is assembled, and pre-synthesized hardware implementations of respective actors are inserted. For actors mapped into software, code is generated, compiled and linked together with other actors into a binary for each processor. Finally, the resulting bitstream is downloaded into the FPGA for rapid prototyping of the final target implementation.

DAEDALUS

Daedalus [145] is another system-level design environment targeted towards streaming, multimedia-type applications. Deadalus is a joint project between the University of Amsterdam and Leiden University in the Netherlands. It combines several tools under a common, XML-based infrastructure to provide application capture, modeling and simulation, and backend platform synthesis

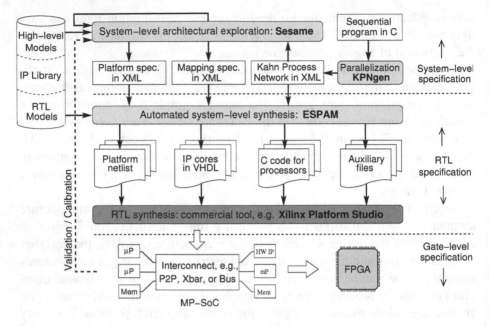

FIGURE 8.3 Daedalus tool flow

functionality (Figure 8.3). At its input, Daedalus accepts applications modeled in a KPN MoC (see Section 3.1.1) that is represented in an XML format. In addition, through a tool called KPNgen, Daedalus can perform automatic conversion of a well-defined subset of sequential C descriptions into a parallelized KPN suitable for input into the Daedalus design flow.

Daedalus supports target architectures consisting of multiple programmable processors and pre-defined hardware IPs. IP components are stored in a library that contains both high-level, functional as well as RTL component models. Given an input KPN and an IP library, a modeling and simulation tool called Sesame allows the designer to assemble a target architecture and perform a mapping of KPN processes onto architectural components. In case multiple processes are mapped to the same processor, Sesame will try to statically schedule processes or insert a lightweight OS kernel. For performance evaluation purposes, processor and IP models in the component library are annotated with tables of estimated execution latencies for typical function-level operations. Sesame links KPN processes to operational latencies of library components they are mapped to. As a result, Sesame will automatically generate a high-level, timed simulation TLM of the specified platform for quick evaluation of selected candidate target architectures. Sesame also allows integration of low-level component models such as cycle-accurate ISSes into the simulation environment. Furthermore, Sesame supports optional automation of the design space exploration process through analytical design space pruning and heuristic search methods such as genetic algorithms.

Given a KPN application, a platform architecture specification and an application-to-architecture mapping (all in XLM form), a final backend synthesis tool called ESPAM automatically generates a description of the selected system implementation. Pre-defined RTL models of all hardware IPs are pulled out of the component library and C code is generated for all KPN processes mapped to programmable processors. Finally, code for each processor is compiled and a hardware models are assembled into a system VHDL model for further synthesis, download and prototyping on an FPGA platform.

PEACE

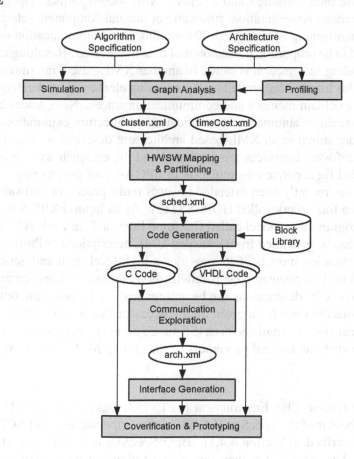

FIGURE 8.4 PeaCE tool flow

PeaCE (Ptolemy extension as a Codesign Environment) [83] is yet another hardware/software co-design framework targeted towards multimedia applications. As the name implies, it is based on Ptolemy [28] as the framework for modeling applications. Ptolemy is a general framework for composition and co-simulation of a wide variety of heterogeneous MoCs in a hierarchical fashion. However, of the many MoCs supported in Ptolemy, PeaCE only accepts combinations of extended SDF and FSM models at its input.

PeaCE realizes a codesign flow from specification over system synthesis down to system prototyping in several steps (Figure 8.4). In a first step, the Ptolemy application model is translated into C code for functional simulation at the specification level. In addition, given a user-defined architecture template consisting of a list of processing elements, performance estimates of application tasks are obtained by profiling each functional block on an ISS of each processor. Annotated application and architecture specifications entered through the user interface are then translated into a generic XML-based format. Operating on this intermediate representation, automatic or manual component selection and HW/SW partitioning is performed. During this step, communication overhead is assumed to be proportional to amount of data transferred. Resulting mapping and scheduling information is stored in another XML-based, intermediate file. Based on this information, code for all processing elements is generated and co-simulated to obtain memory and communication traces. Next, traces are used to drive manual or automatic communication architecture exploration, results of which are stored in an XML-based architecture description. Finally, hardware and software interfaces are generated and the complete system platform is assembled for accurate co-simulation or FPGA-based prototyping.

PeaCE has recently been extended towards multi-processor software development in a framework called HOPES [115]. At its input, HOPES supports a parallel programming model called Common Intermediate Code (CIC), where CIC code can be generated from extended UML descriptions or Ptolemy-based PeaCE application models. CIC provides a high-level, rich and generic API for control or data-oriented code parallelization and inter-process communication. Generic CIC descriptions can be automatically translated into optimized, platform-specific code for a given multi-processor target architecture. Generated code can then be simulated in an ISS-based virtual prototyping environment or downloaded into the real processors of the chosen MPSoC platform.

SCE

The System-on-Chip Environment (SCE) [52] was developed at UC Irvine as the successor of the SpecSyn [64, 63] tool set (the successor of SCE, called ESE, is described in Section 8.4.1). Both SpecSyn and SCE support a PSM MoC (see Section 3.1.2) at their inputs and follow a Specify-Explore-Refine methodology (see Section 1.3). SpecSyn is based on a PSM extension of VHDL called SpecCharts [185]. In contrast, SCE uses the C-based SpecC SLDL (see Section 3.2.3) as the basis for describing all design models throughout the complete design flow. The SpecC language and technology has been standardized

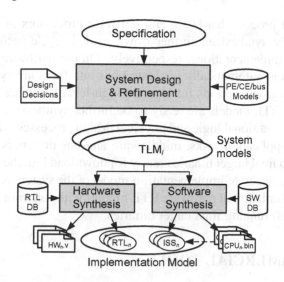

FIGURE 8.5 SCE tool flow

[51] and a derivative of the SCE system-level design frontend has been commercialized and integrated into a complete SpecC-based ESL design solution commissioned by the Japanese Aerospace Exploration Agency (JAXA) [73].

As shown in Figure 8.5, SCE consists of a system design frontend and hardware/software synthesis backend. The design process starts with an abstract specification of the desired system functionality written in SpecC PSM form. In the interactive frontend, the specification is automatically compiled down onto a user-defined MPSoC architecture through a series of architecture, scheduling, network and communication exploration and synthesis steps.

Design decisions such as allocation of architecture components out of the PE, CE and bus databases, scheduling of processes, and mapping of specification processes and channels onto allocated PEs, CEs and buses are entered by the designer through a scripting or graphical user interface. To aid the user in the exploration process, SCE includes retargetable profiling and estimation tools that provide feedback about specification characteristics and effects of decisions on design quality metrics. In addition, SCE supports a plugin mechanism for inclusion of optimizing algorithms that perform automated decision-making.

At its output, the SCE frontend automatically generates TLMs of the system design at successively lower levels of abstractions following a gradual, stepwise refinement processes. Automatically generated TLMs integrate high-level performance models with timing-annotated processes running on top of abstract OS and processor models to provide fast yet accurate analysis and design validation without the need for slow instruction-set simulation.

In a backend process, hardware and software processors in the TLMs are then individually synthesized further down to their cycle-accurate RTL and instruction set implementations, respectively. On the hardware side, application processes and automatically generated bus interfaces are synthesized into VHDL or Verilog descriptions following a high-level hardware synthesis process. Resulting RTL models are ready to be further synthesized and manufactured following traditional logic and physical design processes. On the software side, code for application tasks, middleware and bus drivers is automatically synthesized into final target binaries ready for download into the processors. In addition, a cycle-accurate implementation model of the system is generated that allows for co-simulation of hardware RTL models with software instruction-set simulators (ISSs) running final target binaries.

8.1.2   COMMERCIAL TOOLS

COFLUENT

CoFluent Studio by CoFluent Design [43] is a commercial spin-off based on the MCSE methodology (Méthodologie de Conception des Systèmes Electroniques, also known as CoMES, Co-design Methodology for Electronic Systems) and tool set originally developed at the University of Nantes in France [33]. CoFluent studio is a modeling and simulation environment for early, high-level design space exploration. As a graphical frontend for SystemC, it allows capturing of application functionality, system architecture and their mapping. Application models are specified as networks of timed processes. Processes are described purely by annotated delay estimates, by their functionality given in the form of C, C++ or SystemC code, or as a combination of both timing and behavior. Processes communicate through high-level, message-passing channels, queues, events and shared variables that can also be annotated with estimated communication latencies. The resulting application model can be simulated for early functional and performance evaluation supported by a rich set of built-in graphical analysis and visualization capabilities.

In the next step, an architecture platform can be graphically defined and assembled out of generic processing element or interconnect components. Through drag-and-drop, the designer can map application elements onto the specified platform, and CoFluent studio will generate a SystemC TLM of the resulting architecture for simulation and virtual prototyping. In contrast to other approaches (see below), no detailed component, ISS or bus models are employed. Instead, computation and communication remains at a high level, described as time-annotated processes and message-passing transactions. CoFluent Studio does, however, insert network-level models of communication stacks and OS models for dynamic scheduling of processes mapped onto software processors. All combined, this allows for fast timed simulation at early stages of the design process (at the expense of reduced accuracy).

SPACE CODESIGN

Space Codesign is a recent startup coming out of the École Polytechnique de Montréal in Canada [170]. Its main product is SpaceStudio, which provides a SystemC-based system-level integrated development environment (IDE) built on top of Eclipse (see Section 8.2.1). A specific focus of Space Codesign is support for the increasingly important embedded software development process. Designers can create process-based SystemC application models out of pre-defined library blocks or by importing and wrapping existing C, C++ or SystemC code, where application processes can communicate through message-passing or shared memory channels. Next, a system architecture can be graphically assembled and the application can be partitioned by dragging application blocks onto previously allocated hardware or software processors. As a result, SpaceStudio (through a tool called Elix) will generate a SystemC TLM of the chosen platform where timing-annotated processes are grouped into bus-functional processor blocks, integrated with an OS simulation and connected through register- and cycle-accurate bus models.

All SystemC application models and TLMs generated through SpaceStudio can be simulated for analysis and performance evaluation. High-level models are based on native, host-compiled execution of application processes for fast simulation. In addition, a tool called Simtek will allow creation of cycle-accurate, transaction-level virtual platforms by replacing host simulation of software processes with processor ISS models. Finally, a tool called GenX will take virtual platform TLMs create through SpaceStudio and synthesize them down to a Xilinx FPGA prototyping platform. Software processes are compiled for the selected processor and linked against the target RTOS. Hardware IPs are replaced with pre-designed RTL descriptions, and custom hardware blocks are synthesized using third-party high-level synthesis tools such as Mentor Catapult or Forte Cynthesizer (see Section 8.3.2). Finally, components are assembled into a system netlist for input to the Xilinx FPGA platform synthesis process.

COWARE

CoWare technology started originally as a project at IMEC in Belgium to develop a process-based system-level modeling framework [188]. Since its commercialization, CoWare has evolved into a suite of products that provide a frontend for SystemC TLM capture, modeling and simulation [46]. At the core of the product portfolio, the CoWare Platform Architect is a graphical environment for capturing and assembling virtual system platform models at the cycle-approximate implementation level. CoWare includes an extensive library of detailed component models for hardware IPs, programmable processors and system buses. Hardware IPs are provided either in RTL or bus-functional behavioral form. For programmable processors, ISS models are employed. With the acquisition of LISATek [91], CoWare gained the capability to design application-specific and configurable embedded processors, including generation of associated custom ISSes and software tool chains. Different component models are integrated under a standard SystemC TLM framework using register- and protocol-accurate transactional interconnect models.

Virtual platform models captured and assembled through CoWare's Platform Architect and associated Model Library and Processor Designer can then be simulated using CoWare's own performance-optimized SystemC simulation kernel. Platform Architect thereby includes advanced capabilities for debugging, visualization and analysis of simulation results in order to aid the designer in the overall exploration, platform design and embedded software development process.

SOC DESIGNER

Carbon's SoC Designer [37] is another tool for platform architecture capture and modeling that dates back to simulation technology originally developed at the University of Aachen in Germany. Initially, this technology was marketed under the product name MaxSim by a spin-off called AXYS. Later on, AXYS got acquired by ARM and MaxSim was renamed to ARM RealView SoC Designer. Eventually, ARM sold the SoC Designer product family to Carbon Design Systems.

Similar to other virtual platform tools, SoC Designer includes a graphical user interface for assembling of system architectures out of pre-defined library or user-made custom components. SoC Designer integrates cycle-accurate hardware, ISS and bus models in a proprietary simulation setup. To avoid the need for expensive context switches necessary in typical event-driven SLDL or HDL simulations, components are statically scheduled into a single-threaded, straight-line C/C++ executable that calls individual blocks cycle-by-cycle in a round-robin fashion. This allows for fast yet fully cycle-accurate system simulation. However, components need to be modeled in a specific cycle-callable fashion. SoC Designer includes a frontend for component model development. In addition, existing SystemC, Matlab and VHDL or Verilog RTL models can either be integrated into or co-simulated with the SoC Designer simulation framework.

VAST AND VIRTUTECH

In contrast to virtual platform approaches based on standardized modeling backplanes and languages such as the SystemC, both VaST [187] and Virtutech

[189] are providers of software-centric virtual prototyping solutions based on proprietary simulation technologies. To achieve faster simulation speeds compared to an interpreted ISS, such approaches are based on binary translation or compiled instruction-set simulation of software code. In all cases, simulations are functionally accurate but techniques can vary in terms of achievable simulation bandwidth and cycle-approximate timing accuracy.

Both VaST and Virtutech include graphical environments (called CoMET and Simics, respectively) to integrate software simulators with models of peripherals and other hardware in order to provide full system simulation. In contrast to event-driven system simulation in typical SLDLs, hardware models are directly integrated into the software execution loop, reducing the need for context switches and further speeding up simulations. However, this requires proprietary models to be developed for each hardware block or peripheral. While both companies provide a large library of standard components and graphical frontends to aid in component model development, recent extensions include support for integration of standard SystemC models in such virtual prototyping environments.

On top of virtual prototypes of the platform hardware created with VaST or Virtutech tools, embedded software can then be developed and validated. Both approaches include corresponding software development environments coupled with extensive debugging capabilities (called METeor and Hindsight).

8.1.3 OUTLOOK

In recent years, ESL design concepts, methods and methodologies have experienced increasing interest and adoption in industry. This trend has been accompanied by a growing number of commercially available tools mainly aimed at modeling, simulation and virtual prototyping of complete system platforms and architectures. As technologies mature, we can expect that more and more of the advanced synthesis and design automation solutions currently under development in academia will be transfered into such commercial settings. On the one hand, as described in the following chapters, tools are already emerging that can provide an automated path to implementation from such system-level virtual platform models. On the other hand, additional research and development efforts will be necessary to provide future tools for automation of the design and design space exploration process at the system level. Only automation of the ESL design flow from specification down to implementation will provide the necessary productivity gains that will enable us in the future to close the gap between continously increasing application complexities and exponentiallly growing technological and device-level capabilities.

8.2   EMBEDDED SOFTWARE DESIGN TOOLS

The close relation between embedded software and the underlying customized hardware platform demands special procedures when developing embedded software, for example in terms of: cross compiling, host/target debugging, and testing. With specialized hardware, the embedded software development also needs to take measures for system booting and hardware specific functionality such as system diagnostics and analysis. By its nature, embedded software design has to deal with hardware-specific tools, such as processor specific instruction set simulators, hardware simulators and emulators, and distributed debuggers. This hardware dependency necessitates the use of special development tools.

To aid the development process, hardware vendors provide development environments geared toward their products. For example, the processor IP vendor ARM, provides RVDS (RealView Development Suite) for developing software for various platforms based on ARM cores. The suite integrates ARM cross compilers, enhanced debug capabilities, ARM specific code optimization options, and libraries for common devices (such as flash devices). Similarly, RTOS vendors offer development support tools. Examples include the Tornado tool suite from WindRiver, and MULTI, the integrated development environment from GreenHills. Such development environments are typically point solutions supporting a fixed system architecture. They are less applicable in a scenario in which the target platform remains flexible until the final stage of system design (e.g. complex multi-processor systems), and which may be composed out of heterogeneous components.

Many programmable logic device vendors also provide an embedded software design tool as a part of their design environment. The SOPC Builder from Altera is an example of this, as is the Embedded Development Kit (EDK) from Xilinx. Both of these tools let system designers define and implement a custom platform out of standard building blocks and user defined hardware components. Once the developer has defined and implemented the platform, these tools synthesize the hardware and produce custom software libraries (e.g. for accessing a programmable interrupt controller) reflecting the target's hardware configuration. By generating customized libraries, embedded software design tools like the SOPC Builder and EDK provide some level of abstraction above the hardware (e.g. resolving addressing and basic device access). However, the designer has to manually develop the embedded software on top of the provided low-level primitives for basic device access. Common to both examples is the focus on the vendor specifics of the target platform in terms of processor and RTOS selection. For example, Altera currently supports the NIOS processor with uC/OS-II, whereas Xilinx supports PPC and Microblaze with Xilkernel.

In this way, these vendor-supplied tools are point solutions, that help developers only in case of matching target platforms.

In addition to development tools, simulation environments are important for development of customized embedded systems as development of a hardware prototype is time consuming and a parallel development of hardware and software is desired. HW/SW co-simulation is one approach that allows for an overlapped development of software and hardware, as the SW can be developed on top of a virtual prototype of the hardware. The nature of a suitable approach for simulation highly depends on the envisioned platform complexity, the desired amount of observable simulation features, the required prototype's equivalence to the final software code, and in the needed simulation speed. For simple single core architectures, using an instruction set simulator or processor emulator may suffice. Similarly, for a system that uses one homogeneous RTOS type and does not feature complicated HW interaction, a minimal model may be sufficient, such as a host-compiled RTOS. In a host-compiled RTOS, a modified version of the target RTOS, together with the developed application, is compiled to run on top of host operating system. However, performance limitations make simple solutions such as these infeasible for complex multi-processor SoCs.

In summary, there are many different tools and methodologies currently available for designers to use in developing embedded software. However, these tools are typically point solutions, specific to a vendor or platform. In addition, current techniques rely on the manual development of software. To achieve higher design productivity, a more global approach is desirable, one that can target a wide range of platforms and has, furthermore, a path to synthesis.

Next, we will outline some academic and commercially available tools for embedded software development and generation.

8.2.1 ACADEMIC TOOLS

ECLIPSE

The open source Eclipse [59], is a multi-language software development platform. It consists of an Integrated Development Environment (IDE) with a flexible plug-in system. The IDE provides a source code editor with a rich set of source annotation and browsing capabilities, integrates a compiler, a source code debugger and many more facilities to aid the software development process. Eclipse's primary focus is the Java language, hoverer with various plug-ins it addresses many other languages as well, such as C/C++, Cobol, Python, Perl, PHP. Eclipse's well defined plug in system makes it very attractive for customized extensions.

With the popularity of Eclipse IDE, many academic and commercial providers use Eclipse as a platform for their own products with a wide range of specific functionalities. For example, plug-ins exist for UML-based capturing and development (e.g. IBM Telelogic Rhapsody [94]). They extend the IDE with an interface to graphically capture UML-diagrams and later generate structural source code (e.g. class hierarchy) out of the diagrams. Many Eclipse plug-ins more specifically target embedded software development. One example is the Tensilica Xtensa Xplorer IDE [97]. It provides a GUI for customizing an Xtensa processor, integrates a specific cross compilation tool chain and furthermore offers co-simulation and emulation integration. Another Eclipse plug-in example addresses automotive software component design following the AUTOSAR standard, Greensys' Autosar Builder [67]. It supports developing AUTOSAR Software Component (SW-C), ECU and System descriptions at the applications level, integrates their validation and end emulation. Many more plug-ins exist, which we can not enumerate there. The wide range of highly specialized plug-ins make Eclipse an very versatile and powerful software development environment.

POLIS

The POLIS system [11] developed at UC Berkeley is a hardware/software co-design environment with a focus on reactive systems. POLIS allows the user to specify the application in a high level language such as the Esterel or using a graphical as FSM notation. The input specification is internally converted into a co-design finite state machine (CFSM) model. Each FSM within a CFSM represents a component in the system. Using this CFSM, POLIS allows the designer to partition the design, formally verify it, co-simulate as well as synthesize portions of the system. Software generation is performed by transforming the CFSM sub-network chosen for SW implementation into an S-Graph, and subsequent C code generation. In addition an application specific scheduler and drivers are generated for each partitioned design.

DESCARTES

DESCARTES [162] is a software synthesis environment that targets real-time signal processing systems. It focuses specially on optimization techniques for mapping data flow oriented block diagrams onto a DSP. It provides a combination of different mapping and optimization strategies that allow comfortable synthesis of real-time code which is highly adapted to application-specific needs as imposed by constraints on memory consumption, sampling rate, or latency.

DESCARTES uses a data flow description (Asynchronous Data Flow (ADF) and an extended Synchronous Data Flow (SDF)) as an input. The work provides scheduling algorithms defining the order of execution for each computation kernel (node) in the data flow following input constraints of latency, throughput and memory consumption. It generates C code for each computation kernel that then is compiled using a DSP specific C compiler. With the choice of input model, DESCARTES is tightly coupled to the signal processing domain. In contrast, a flexible generic C-programming model is desirable over these specific input models to cater to the needs of a broader programming audience and to capture a wider range of application domains.

8.2.2 COMMERCIAL TOOLS

MATHWORKS: REAL-TIME WORKSHOP

MathWorks offers a range of packages that are centered around Matlab, a numerical computing environment and programming language. Simulink [132] is a commercial model-based design tool for modeling, simulation and analysis of multi-domain systems. As an input, Simulink has a graphical user interface for assembling a system as a block diagram describing the system functionality. Blocks within Simulink are typically library defined containing standard signal processing (e.g. filters) and control functions. They are connected and hierarchically composed to express the system either as discrete timed or continues timed models. Simulink is tightly integrated into the Matlab environment, and widely used for simulation and design in the control theory and the digital signal processing domain.

On top of Simulink, MathWorks offers Real-Time Workshop [131] for the synthesis of an software implementation. It generates stand-alone C code for algorithms modeled in Simulink. The generated code can be used in many real-time and non-real-time applications, as well as for simulation acceleration and hardware-in-the-loop testing. Real-Time Workshop generates ANSI/ISO C or C++ code from discrete, continuous, or hybrid Simulink models for execution on a wide range of target platforms. It can target bare processors without any operating system, as well as multi-tasking systems with an RTOS.

DSPACE: TARGETLINK

TargetLink [53] is a code generator, by dSpace. It integrates into the Matlab/Simulink environment and is similar to the above discussed Real-Time Workshop. It uses Matlab/Simulink as a graphical editor for system capture. However, it comes with an own library of block components for graphical design composition.

TargetLink provides generation of production code out of a Matlab/Simulink model for a wide range of target processors and platforms. TargetLink mainly addresses the design of automotive systems. It supports targeting OSEK/VDX-compliant operating systems [92] for integration of the generated function code onto an Electronic Control Unit (ECU).

dSpace offers both hardware and software solutions for the automotive design. For validation and testing of applications, it provides three levels of model testing. Model-in-the-Loop (MiL) executes the original model, validating functionality and dimensioning of the algorithm. Software-in-the-Loop (SiL) executes the generated software code on the simulation host, for validation of the implementation. Hardware-in-the-Loop (HiL) executes the final software on an actual ECU. The inputs and outputs of the ECU are controlled by a Matlab/Simulink model simulating the physical control environment.

In summary, dSpace TargetLink, offers a comprehensive solution for the design, synthesis and test of automotive designs with a focus on software. Current development extensions are addressing the emerging AUTOSAR [9] as multi-core ECU platforms.

ESTEREL TECHNOLOGIES: SCADE

Esterel Technologies' commercial SCADE suite [57] is a development environment for system and software engineers targeted for safety-critical applications. With its editor complex systems are captured using a graphical notation for hierarchical composition of data flow and safe state machine notations. SCADE comes with a rich library of predefined blocks for operators, linear functions, digital functions, filters, state machines and model composition. The product is internally based on the synchronous data-flow programming language Lustre [85]. The tool suite is mainly used in the aerospace and defense domains.

SCADE offers a C code generator (KCG) that is certified for the development of airborne systems and equipment, which allow the production use of the generated code. The code generator translates each block of the system specification into a software implementation that can be integrated for execution on a target processor.

For the analysis of generated code, SCADES integrates with external tools for Worst Case Execution Time (WCET) and stack utilization analysis. They provide WCET and stack utilization information at the model level, detailed for each function block within the specification. These analysis capabilities, provide design quality feedback about maintaining timely execution and staying withing resource constraints, which are important for safety critical systems early in the process supporting an efficient design.

In addition, Esterel Technologies offers gateway integration with other modeling environments that allow importing specifications and requirements into SCADES. For example, it provides a gateway for importing of discrete controllers prototyped in Matlab/Simulink. It further integrates with Rhapsody UML/SysML for high-level system requirements. These gateways expand the coverage of SCADES tool suite to other modeling approaches.

UML/SYSML PRODUCTS

The Unified Modeling Language (UML) [147] is an standardized language for the specification of software systems. It is a language for specifying, visualizing, constructing, and documenting the artifacts of a system with an emphasis on the earliest part of a design process. UML is a modeling language, in contrast to a a programming language. It therefore focuses on capturing relevant information required for understanding the design problem, solving it, and guiding implementation of the solution. It excludes any irrelevant information that may hinder that progress.

UML defines 13 different datagram types with a wide range of modeling system structure, system behavior and the interaction of system elements. With this range of diagram styles it is apparent that the designer has great flexibility in capturing system structure. UML provides means to capture boundaries, requirements and system interaction. On the other hand UML by itself is not very suitable to concisely express formulas. For capturing algorithms in the system, UML often relies on embedded C, C++, or Java code as a description.

The Systems Modeling Language (SysML) [149] an extension of a subset UML by using UML's profile mechanism. SysML reduces UML's restriction to software-centric systems, and is positioned as a modeling language for systems engineering applications. It only uses 7 out of the 13 UML diagrams, but extends it by additional diagrams and concepts. For example, it adds requirement diagrams for capturing parametric constraints between structural elements, which aid performance and quantitative analysis. It also introduces additional MoCs by extending the behavior of UML activities for the modeling of continuous and probabilistic systems. The use of UML and SysML for system level design of SoCs is discussed in [116, 126].

Many commercial products for model-based development exist, which are based on UML/SysML. Examples include IBM Telelogic Rhapsody [94], Spark Systems' Enterprise Architect [175], Gentleware's Poseidon for UML [69] and Artisan Software's Artisan Studio [169]. These tools offer graphical editors for capturing UML/SysML diagrams, the analysis and consistency validation. In addition these tools offer generation of targeted code for framework integration. The framework code itself may not contain all algorithm code, however provides a start framework for manual software development.

8.2.3   OUTLOOK

With the increasing attention to embedded software design, the tool support for developing embedded applications has significantly improved in the recent years. Vendors of hardware (e.g. FPGA) and software products (e.g. RTOS) provide an added value to their products by offering integrated development environments with specialized support for their own product. In addition, many domain specific specialized solutions guide the application development for example in the automotive and signal processing domain. A stronger focus on better structured, reusable, and expandable software implementations is noticeable, for example through utilizing component-based principles such as in AUTOSAR or through tighter connecting documentation and implementation as seen in an UML-based process.

The complexities of future platforms will continue to grow. We will see systems with diverse distributed heterogeneous components as well as systems with many cores. As platform complexities grow, manually implementing embedded software will become infeasible, especially when considering the decreasing time-to-market. Therefore, there is an essential need to further simplify the modeling and development of software and systems. In particular, design environments are needed, which enable abstract development of complex systems at the algorithm level, which automate the implementation process through automatic synthesis of both hardware and software, and which allow the designer to focus on essential functional aspects without the burden of low-level implementation details.

8.3 HARDWARE DESIGN TOOLS

Research and tool development for hardware design-automation began four decades ago, and progressed through four phases. The 1970s embodied the concept phase, which gave birth to basic definitions for the languages, design methods, and tools necessary for standard and custom processors. The 1980s introduced the algorithm phase, which saw a flurry of research activities defining algorithms for allocation, binding, and scheduling in a new field called High-Level Synthesis (HLS). During the decade which followed, these new approaches were consolidated with the emergence of several seminal books on HLS and the first commercial tools. Finally, the first decade of this century ushered in the acceptance phase, during which the concept of automatically generating custom hardware components from high-level programming languages (C-to-RTL) has become accepted and applied to many custom designs by industrial designers world-wide.

The concept phase began with Bell and Newell's seminal book on computer structures [16], which introduced Instruction-Set Processor (ISP) notation. ISP was intended to precisely and unambiguously describe the behavior of instruction-set processors. This behavior was characterized by the existence of an interpretation algorithm that fetches, decodes, and executes "instructions" stored in the memory. The ISP concept was refined by Barbacci at CMU who introduced the Instruction-Set Processor Specification (ISPS) for the simulation, evaluation, and synthesis of simple processors [14, 15, 13]. Barbacci, along with Siewiorek, also developed an initial system for the synthesis of processors called CMU RT-CAD System in 1976 [168]. That opened broader investigations into the different aspects of synthesis process such as internal representations [133], component allocation [84] and processor architecture selection [179]. At this same time, Zimmermann and Marwedel at Kiel developed the MIMOLA design method to design of digital processors from a very high-level behavioral specification [199, 127]. A key feature of this method is the synthesis from application programs expected to run on that processor. This was the first attempt at C-to-RTL compilation.

In the 1980s, research on algorithms for HLS spread to many different countries. This research was focused on languages and representations, algorithms and methodologies, and tools and environments. In terms of languages every research group used a different subset since standard languages such as C or VHDL were not synthesizable [122]. In the representation domain CDFG became popular at this time [151]. Allocation, binding and scheduling algorithms were the most popular topic for research [155, 128, 10, 34, 49, 156, 153]. This was a time of great diffusion of new ideas. Different methodologies for the design of controllers, datapaths or complete custom processors were introduced based on different design paradigms [26, 50, 152, 176, 177, 181, 153, 134]. Similarly, many HLS tools came into use, the most prominent being the Yorktown Silicon Compiler from IBM [25] which included high-level, logic and layout synthesis, CATHEDRAL from IMEC in Belgium [160], which focused on multiprocessor DSP applications, as well as The System Architect's Workbench from CMU [176], and Design Environment from U of Karlsruhe [36].

The consolidation phase of HLS in 1990s is characterized by the appearance of several books defining the seminal work in the field. Don Thomas and associates published a book on CMU's System Architect's Workbench in 1990 [178], followed by several other books by different authors on different aspects of HLS, including timing constraints [114], methodologies and algorithms [61], digital signal processing [186], synthesis and optimization [139], and component reuse [102]. Several edited books concerning the issues involved in HLS [35, 138] were also published in that period. 1990s were also characterized by the appearance of EDA companies offering commercial tools. Wakabayashi introduced NECs Cyber synthesis tool [191], Synopsys introduced Behavioral Compiler (BC) [110], and Mentor introduced Monet [56]. Those early tools followed basic principles of HLS as described in the above mentioned HLS books. For example, BC accepted a behavioral description in a subset of VHDL or Verilog. It converted the input description into a CDFG representation that exposed control and data dependences. In order to perform technology-specific scheduling BC converted data flow graph in each basic block of CDFG into gates in order to produce accurate delay estimates. This way BC could schedule two operations into the same clock cycle as long as their joint delay was smaller than the clock cycle. After scheduling, BC performed allocation and binding and synthesized the control FSM with gates. The last step was logic optimization of the generated datapath and controller.

The early tools showed the possibility of HLS automation. However, there were several obstacles for commercial success. Designers had to use a tool-dependent subset of HDLs instead of a standard programming language such as C or Java. Datapath and controller architectures were overly simple without pipelining or data forwarding. The controller was implemented as an FSM with gates, so that later upgrade or changes needed re-synthesis. Since the controller did not use control or program memory, it was not possible to execute large programs. Even when the synthesized result was acceptable, interfacing the synthesized component into a larger system was not well defined.

The largest obstacle to widespread acceptance of HLS was the market's unpreparedness for processor-level abstraction. This has changed dramatically in this decade because of increased system complexities. The new HLS tools use a standard programming language as the input and generate RTL in a HDL as the output so synthesized designs can be prototyped with FPGA tools. Moreover, the quality of these tools has improved through the use of more sophisticated algorithms. At the same time the complexity of synthesized components increased from special functions with a FSM controller to custom processors with a programmable controller.

8.3.1 ACADEMIC TOOLS

GAUT

The GAUT tool from UBS [157] is an academic and open-source HLS tool dedicated to digital signal processing applications. It generates an independent custom processor with custom interface that allows it to be inserted into any system. Starting from an algorithmic bit-accurate specification written in C/C++, GAUT extracts the potential parallelism before performing the allocation, scheduling and binding tasks. The mandatory synthesis constraints are the throughput, the clock period, and the target technology while the optional design constraints are I/O timing diagram and the variables-to-memory mappings. GAUT synthesizes a potentially pipelined architecture composed of a processing unit, a memory unit, a communication interface unit that uses a globally-asynchronous, locally-synchronous protocol.

GAUT generates an IEEE P1076 compliant RTL level VHDL file. This VHDL file is an input for commercial, off the shelf, logical synthesis tools such as ISE/Foundation from Xilinx, Quartus from Altera, or Design Compiler from

Synopsys. GAUT also generates a SystemC cycle-accurate simulation model for simulation-based validation.

NO-INSTRUCTION-SET COMPUTER

The No-Instruction-Set Computer (NISC) from UCI [40] is an attempt to overcome two of the weaknesses of HLS: programmability and metric closure. Most HLS designs are special function components with a fixed controller that implements the FSM of the special function executed in the datapath. Such a controller is usually implemented with gates which limit the FSM size to a couple of hundred states. The first problem with such an implementation is that the complete design has to be re-synthesized for any change or upgrade in the given function. The other problem is that this type of implementation can not support large amounts of code. To solve this problem NISC uses a programmable controller with a control-word memory that stores control words for every clock cycle. This way large codes can be accommodated and even dynamically up-loaded.

The second HLS weakness is that during synthesis and optimization the required metrics must be estimated. The exact value of delay, power, and performance is not known until the final layout. The finalized metrics values or metric closure is needed to fine tune the architecture and the application code. NISC solves this problem by separating the allocation and datapath structure generation from scheduling and binding performed by the NISC compiler. Therefore, making it possible to a create complete structure with all the metrics known before compilation. If the final results are not acceptable, the datapath can be modified and the application code recompiled. Furthermore, NISC methodology leads to the concept of standard architecture-cells or templates that can be stored in the library and used by different application designers. Having several such templates per application domain greatly simplifies the methodology and tools on lower levels of abstraction.

A NISC tool set as shown in Figure 8.6 consists of three different components: a datapath generator, a NISC Compiler, and an RTL Generator. The datapath generator is used to create a datapath structure for a given application. This task can be done automatically by profiling the application code in C, compiling usage statistics, selecting components and connectivity for the given performance metrics and generating a Generic Netlist Representation (GNR) of the datapath. A datapath template can be also selected from the template library, or designers can specify their own datapath by creating a GNR through GUI. The NISC cycle-accurate compiler [161] compiles the application for a given datapath. It converts the application code into a control-words stream controlling datapath on each clock cycle. The RTL Generator produces the RTL description for inputing to FPGA or ASIC tools. It converts the datapath and controller GNR into RTL with control words generated by the compiler loaded into the control-word memory in the controller.

If synthesized results are not satisfactory, the datapath structure and/or application code can be modified. This can be done manually by rewriting the application code and GNR or automatically through code refinement or datapath refinement tools.

A NISC enables the designers to control every aspect of the design. The designer can select the exact points for improvement and then make the changes quickly. For example, by changing the GNR description of the datapath architecture, the designer can reduce a critical path delay or fix complex multiplexers and connections that consume too much power or make the layout unroutable. Since datapath can be an input in NISC technology, the designer can selectively explore options for quality metrics. For example, a designer can focus on dynamic power minimization by modifying the connections or gating or latching them in the datapath description and quickly see the effect on the final results.

SPARK HIGH LEVEL SYNTHESIS

SPARK tool from UCSD [140] is a C-to-VHDL high-level synthesis framework that employs a set of innovative compiler, and synthesis transformations to improve the quality of high-level synthesis results. The SPARK parallelizing high-level synthesis methodology is targeted particularly to multimedia and image processing applications along with control-intensive microprocessor functional blocks.

As shown in Figure 8.7, SPARK takes a behavioral description in ANSI-C as input. It also takes additional information as input, such as a hardware

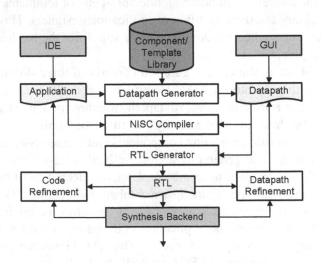

FIGURE 8.6 NISC technology tools

resource library, resource and timing constraints and user directives for the various heuristics and transformations. SPARK stores the input behavior in a hierarchical intermediate representation, a CDFG derivative with dependences across basic blocs. This is critical for enabling coarse-level transformations and making global decisions about code motion.

SPARK first applies a set of coarse-grain and fine-grain code transformations to the input description during a pre-synthesis phase before performing the traditional high-level synthesis tasks of scheduling, allocation and binding. The transformations in the pre-synthesis phase include (a) coarse-level code restructuring by function inlining and loop transformations, (b) transformations that remove unnecessary and redundant operations such as common sub-expression elimination (CSE), copy propagation, and dead code elimination (c) transformations such as loop-invariant code motion, induction variable analysis (IVA) and operation strength reduction, which reduce the number of operations within loops and replace expensive operations with simpler operations.

The pre-synthesis phase is followed by the scheduling and allocation phase. Resource allocation and module selection are done by the designer and are given as input to the synthesis tool through a hardware resource library. The scheduler is organized into two parts: the heuristics that perform scheduling

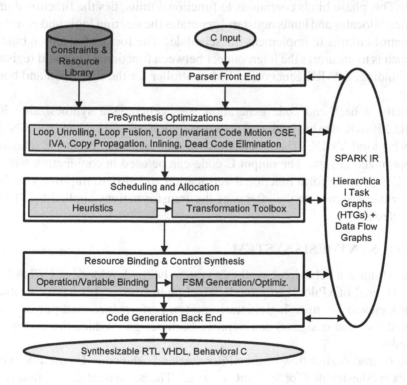

FIGURE 8.7 The SPARK Synthesis Methodology

and a transformations toolbox. The transformations toolbox contains speculative code motion transformations, the percolation and trailblazing code motion techniques, dynamic renaming of variables et cetera. The synthesis transformations include chaining operations across conditional blocks, scheduling on multi-cycle operations, and resource sharing.

Besides the traditional high-level synthesis transformations, the scheduling phase also employs several compiler transformations applied "dynamically" during scheduling. These dynamic transformations, such as dynamic CSE and dynamic copy propagation, exploit the new opportunities created by code motions. A branch balancing technique also dynamically adds scheduling steps in conditional branches to enable code motions, specifically those code motions that duplicate operations in conditional branches.

Passes from the toolbox are called by a set of heuristics that guide how the code refinement takes place. The heuristics and the underlying transformations that they use are kept completely independent from each other. This allows the heuristics to employ the various transformations as and when required, thus enabling a modular approach that allows the easy development of new heuristics.

The scheduling phase is followed by a resource binding and control synthesis phase. This phase binds operations to functional units, ties the functional units together, allocates and binds registers, generates the steering logic and generates the control circuits to implement the schedule. The focus of resource binding approach is to minimize the interconnect between functional units and registers. After binding, SPARK generate a FSM controller for the scheduled and bound design.

Finally, a back-end code generation pass generates a synthesizable RTL VHDL. SPARK also has back-end code generation passes that generate ANSI-C and behavioral VHDL. These behavioral output codes represent the scheduled and optimized design. The output C code can be used in conjunction with the input C code to perform functional verification and also, to improve visualization for the designer on the effects of the transformations applied by SPARK on the design.

XPILOT SYNTHESIS SYSTEM

The xPilot is a behavioral synthesis system being developed at UCLA [183, 41]. The goal of xPilot is to provide novel platform-based behavior synthesis technologies to optimize logic, interconnects, performance, and power simultaneously, so that designers can improve both design productivity and quality of results.

The overall design flow of the xPilot system is shown in Figure 8.8. xPilot accepts synthesizable C or SystemC as input. The behavioral description is first parsed and optimized by the UIUC LLVM compiler infrastructure. A System-level Synthesis Data Model (SSDM) is then constructed from the LLVM's internal representation. The basic building blocks in SSDM are processes and channels. A process describes the behavior of one module, and each process uses a CDFG to capture its behavior. Each process interacts with other processes through ports and channels.

Each channel implements some interface to implement certain communication protocols. In total, an SSDM defines a process network to model the concurrent behavior of a complex system. On top of SSDM, xPilot performs platform-based synthesis and physical-aware optimizations during scheduling and resource binding; these construct an optimized State Transition Diagram (STG) and an associated datapath model. At the back end, xPilot generates RTL implementations together with constraint files such as multi-cycle path constraints and physical location constraints, to leverage the existing logic synthesis and physical design toolset.

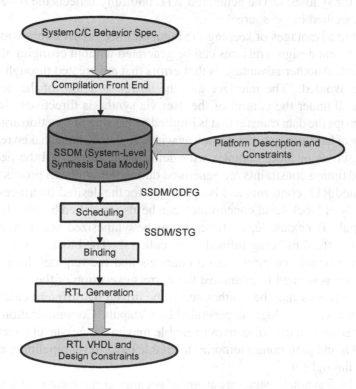

FIGURE 8.8 xPilot Synthesis System

8.3.2   COMMERCIAL TOOLS

CATAPULT SYNTHESIS

Catapult from Mentor [137] takes a behavioral description written in ANSI C++ and a set of user directives as input and generates an RTL that is optimized for the specified target technology. The input specification is behavioral and does not include any notion of explicit parallelism, time, state or interface protocol or the design structure. Required directives specify the selected component library and the clock period. Optional directives control hardware details such as interface and memory mappings, how much parallelism to implement in loop unrolling and loop pipelining, hardware hierarchy and block communication, latency or cycle constraints for scheduling, the number and/or type of hardware resources for allocation , etc. Catapult supports native C++ integer types as well as C++ bit accurate integer and fixed-point datatypes are supported for synthesis. The generated RTL faithfully reflects the bit-accurate behavior specified in the source.

One of the advantages of keeping the input untimed is that a very wide range of interfaces and design structures can be generated without changing the input specification. Another advantage is that errors that are created through manual coding are avoided. The interface and the design structure of the generated design are all under the control of the user via synthesis directives. Interface synthesis maps the data transfer that is implied by passing of function arguments to a variety of hardware interfaces such as wires, registers, handshaked registers, memories, buses or more complex user-defined interfaces. All the necessary signals and timing constraints are generated during the synthesis process so that the generated RTL conforms and is optimized for the desired interfaces.

Hierarchy or block-level concurrency can be also specified by user directives with Catapult. For example, a C function can be synthesized as a separate hardware block instead of being inlined in its caller(s). The blocks are connected with the appropriate communication channels and the required handshaking interfaces are generated to guarantee the correct execution of the specified behavior. The blocks may be synthesized to be driven by different clocks. The clock domain crossing logic is generated by Catapult. Communication is optimized depending on user directives to enable maximal block-level concurrency using FIFOs and ping-pong memories to enable block-level pipelining and thus improved throughput.

All the HLS synthesis steps are aware of accurate component area and timing numbers for the target technology (ASIC or FPGA) for the RTL synthesis tool of choice. Accurate timing and area numbers for components are essential for generating an RTL that meets the timing and area constraints. During synthesis, Catapult queries the component library so that it can allocate a variety of combinational or pipelining components with different performance and area tradeoffs. The queried component library is pre-characterized for the target technology and the target RTL synthesis tool. Component libraries can also be built by the user to incorporate specific characterization for memories, buses, I/O interfaces or other pieces of functionality such as pipelined components..

The synthesis process generates the required verification infrastructure in SystemC so that the input stimuli from the original C++ testbench may be applied to the generated RTL to verify its functionality against the original input specification using simulation. The synthesis process also generates the required verification infrastructure for sequential equivalence checking between the input specification and the generated RTL. Catapult has been successfully used in over 200 ASIC tapeouts and several hundred FPGA designs. Typical applications include computation-intensive algorithms in communications and video and image processing.

CYNTHESIZER

Cynthesizer from Forte [58] takes a SystemC module containing hierarchy, multiple processes, interface protocol and algorithm and produces RTL Verilog optimized to a specific target technology and clock speed. The target technology is specified by a user provided library file or, for FPGA implementation, by identifying the targeted Xilinx or Altera tools.

The input to the high-level synthesis flow used with Cynthesizer is a pin- and protocol-accurate SystemC model. The designer puts untimed high-level C++ in a hardware context using SystemC to represent the hardware elements such as ports, clock edges, structural hierarchy, bit-accurate data types and concurrent processes.

Clocked thread processes are used for the majority of the module functionality. They contain an infinite loop that implements the bulk of the functionality along with the reset code that initializes I/O ports and variables. Within a thread, the designer can combine untimed computation with cycle-accurate communication. A hybrid scheduling approach is used in which the protocol sections are scheduled in a cycle-accurate way, honoring the clock edges specified by the designer as SystemC wait statements. The computation code is written without any wait statements and scheduled by the tool to satisfy latency, pipelining and other constraints given by the designer. Triggered methods can also be used to implement behaviors that are triggered by activity on signals in a sensitivity list, similar to a Verilog 'always' block. This allows a mix of high-level and low-level coding styles to be used if needed.

Complex subsystems are built and verified by combining modules using structural hierarchy just as it would be done in Verilog or VHDL. The high-level models used as the input to synthesis can be simulated directly to validate both the algorithms and the way the algorithm code interacts with the interface protocol code. Multiple modules are simulated together to validate that they interoperate correctly to implement the functionality of the hierarchical subsystem.

In order to ensure that the synthesized RTL meets timing at a given clock rate using a specific foundry and process technology, a high-level synthesis tool requires accurate estimates of the timing characteristics each operation. Cynthesizer uses an internal datapath optimization engine to create a library of gate-level adders, multipliers, etc. The timing and area characteristics of these components are used by Cynthesizer to make tradeoffs and optimize the RTL. Designers have the option of using the gates for implementation or of giving their logic synthesis tool RTL representations of the datapath components.

Cynthesizer produces RTL Verilog for use with logic synthesis tools. The RTL consists of a finite state machine and a set of explicitly instantiated datapath components such as multipliers, adders, and multiplexors. More complex custom datapath components that implement arithmetic expressions used in the design are automatically created, and the user can specify sections of C++ code to be implemented as datapath components. The multiplexors directing the dataflow through the datapath components and registers are controlled by a conventional finite state machine implementation.

SystemC is a good fit for high-level synthesis because it combines the high-level and object-oriented features of C++ with hardware constructs that allow a designer to directly represent structural hierarchy, signals, ports, clock edges etc. This provides a very efficient design and verification flow in which behavioral models of multiple modules can be concurrently simulated to verify their combined algorithm and interface behavior. Most functional errors can be found and eliminated at this high-speed behavioral level instead of through time-consuming RTL simulation. Once the behavior is functionally correct, the models that were simulated are used directly for synthesis, eliminating opportunities for mistakes or misunderstanding.

PICO

PICO tools developed by Synofra [45] support the development of custom processors or application engines for a system platform consisting of standard CPUs and DSPs, memories, IF components such as DMAs or USBs and complex application engines such as video codecs and wireless modems. PICO provides a fully automated, performance-driven, application engine synthesis methodology that enables true algorithmic-level input specification. It produces C-to-RTL mapping under performance constraints in terms of throughput and cycle-time. The key to PICO's approach is the use of an advanced parallelizing compiler in conjunction with an optimized, compile-time configurable architecture template to generate an application-engine RTL.

PICO uses C/C++ language as the preferred mode of input specification at the algorithmic level to allow the user to specify functionality as a sequential program. PICO's parallelizing compiler automatically extracts parallelism from the input specification to meet the desired performance based on its analysis of program dependencies and external resource constraints. PICO is intended for applications that process data streams such as audio, video, imaging, security, wireless, networking applications, among others. There is large amount of parallelism in such applications at various levels of granularity. These applications consist of a sequence of transformations expressed as multiple loop-nests encapsulated in a C procedure that is executed repetitively on a stream of data blocks.

One invocation of this procedure is called a task. PICO optimizes parallelism on task-level, loop-level, iteration-level, and instruction level at the same time to satisfy performance and cost constraints. Given the parallelism available in the application code at various levels, the PICO compiler exploits this parallelism without violating the sequential semantics of the application code by following the well-defined model of Kahn process networks, in which a set of sequential processes communicate via streams through unbounded FIFOs. This Kahn process network concept is implemented in PICO with an architectural template defined by a Pipeline of Processing Arrays (PPA). Each of the top level loop-nests in the C procedure is mapped to a custom processor called Processing Array (PA) which communicates with other PAs via one or more FIFOs or memories. Each PA is structured like a wide Very Long Instruction Word (VLIW) processor that is customized to execute only one program: a loop iteration.

Along with the hardware RTL and its related software, PICO also produces SystemC-based TLM models of the hardware at two levels of abstraction: an untimed programmer's view and a timed programmer's view. Knowledge of the target technology and its design trade-offs is embedded as a part of a macro-cell library which PICO tools use as a database of hardware building blocks. The library consists of pre-verified, parameterized, synthesizable RTL components such as register, adders, multipliers and interconnect elements. These macrocells are independently characterized for various target technologies and various macrocell parameters. PICO uses these characterization data for its internal delay and area estimates.

CYBERWORKBENCH

CyberWorkBench (CWB) from NEC is a C-based high-level synthesis and verification tool that has been in development since 1990s [190, 191, 144]. The main idea behind the CWB is an "all-in-C" approach in which all the modules in the design are described in the behavioral C language. CWB also supports legacy RTL blocks as black boxes, which are called as C functions. At the same time the synthesis, verification, and debugging tasks are all done in the C source code.

CWB targets general SoC platforms which normally contain several CPUs or DSPs, in addition to custom HW modules and some pre-designed or fixed RTL or gate level IP modules that are connected directly or through buses in the platform.

Initially, each custom HW module is described in a specialized behavioral C called Cyber-C. Once its functionality is verified through the C simulator and debugger, the HW module is synthesized with the behavioral synthesizer. The custom processors are also synthesized from their C description in the CWB environment. Legacy RTL blocks are described as functions and handled as black boxes. The CPU bus and other bus interface circuits are also automatically generated using a CPU bus library. After synthesis and verification of each module, the CWB environment allows designers to create a cycle-accurate simulation model for the entire platform including CPUs, DSPs and custom HW modules. With this model designers can verify both the functionality and the performance of their design, as well as the embedded software running on the CPU, DSP and custom processors. The behavioral synthesis is fast enough to allow designers to modify and synthesize HW modules and embedded software many times. The input C code can also be debugged with a formal verification tool that checks properties and model assertions. These global properties and in-context assertions are described in the original input C code. The equivalence between the behavioral C and the generated RTL can be verified dynamically and statically.

Currently, the platform-level parallelization is left to the system designers. They partition the input C code into individual HW modules and embedded software based on the performance results of the cycle simulation or FPGA prototyping.

BLUESPEC

Bluespec tools from Bluespec provide an alternative to the standard C-based HLS technology by focusing on components that do not fall into the loop-and-array paradigm: processors, caches, interconnects, bridges, DMAs, I/O peripherals, and similarly others. These components are characterized by heterogeneous, irregular and complex parallelism for which the sequential computational model of C is not expressive enough. They use a language in which the concurrent behavior of a system is expressed as a collection of rewrite rules. Each rule has a guard expressed by a Boolean predicate on the current state, and an action that transforms the state of the system. These rules can be applied in parallel, that is, any rule whose guard is true can be applied at any time. The only assumption is that each rule is an atomic transaction, meaning that each rule observes and delivers a consistent state, relative to all other rules. The rules and their ordering are described in Bluespec System Verilog (BSV).

BSV allows designers to specify the micro architecture precisely, but with powerful generative and parameterization mechanisms which allow a single source to flexibly represent a family of micro architectures, within which different choices may be appropriate for different metric optimizations. Thus BSV provides synthesis from very high level description with a precisely-specified micro architecture in the parameterized program structure. Bluespec Compiler compiles a BSV description into Verilog RTL or SystemC while Bluspec Simulator simulates Bluespec designs with cycle accuracy.

8.3.3   OUTLOOK

The last thirty years of research and development into high-level synthesis has proven profitable, as evinced by the increasing supply of HLS tools and by designers' acceptance of C-to-RTL concepts. Though there has been much progress in the concepts, algorithms, and methods for HLS, there is more work ahead, which is driving the surge in HLS research and tools [45].

Although some tool suppliers are offering specific languages that support efficient descriptions of functionality or architecture, most of the market is settling on C/C++ for describing input functionality. That decision is leading to increasing efforts in pre-synthesis compilation to increase possible concurrency for future optimization and to improve the quality of the synthesized design.

The synthesized architecture is usually the set of storage and functional-unit components connected through multiplexers. Still, much work must be done to improve the architecture by adding busses, control and datapath pipelining, and programmable controllers in order to move the architecture into direction of custom processors. Some suppliers offer specific pipelined-blocks architecture for "loop-and-array" applications, but there is no conclusion on standard architecture-cells or templates that will make C-to-RTL compilation more efficient, as in the compilation of C to instruction-sets.

Moreover, the problem of interfacing synthesized components and merging them into a system platform is still grossly under solved. As with component architecture, there is a need for standard interface-cells so that any two synthesized components can be easily connected. With availability of architecture and interface standard cells and an efficient compilation from C, the directions of the IP industry in the future still remain to be answered.

8.4   CASE STUDY

So far we have looked at a variety of system level, software and hardware design tools. Many tools are available publicly or commercially to assist with different aspects of embedded system design. We advocate that there will be a need for new tool-sets or design environments that integrate different aspects of embedded system design. These developments will be crucial to the evolution of a model based design and verification methodology for embedded systems. In the long term, there will be no distinction between hardware and software at the design entry stage. The next generation of embedded system design tools will focus on applications and enable non-experts to design embedded systems.

In this section, we will present a case study for the design of an industrial size application, the MP3 decoder. We use the Embedded System Environment (ESE) tool set [39] to present the model based design of the MP3 decoder on four heterogeneous embedded platforms. The ESE tool flow embodies the design methods and principles that have been discussed in this book. We will present results that demonstrate the speed and accuracy of automatically generated models, the quality of the synthesized design and the productivity gains that results from using ESE. The case study is meant to motivate designers to adopt the embedded system design methods and principles presented in this book.

8.4.1   EMBEDDED SYSTEM ENVIRONMENT

ESE consists of two parts, the front end and the back end, as shown in Figure 8.9. The input to front end is the system specification consisting of an application model mapped to a given platform. It automatically generates a TLM of the system for fast and early design evaluation. The back end reads this TLM and synthesizes the required software and hardware to produce the

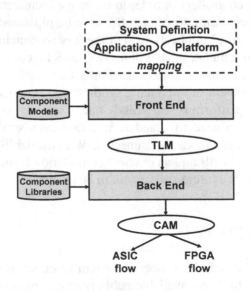

FIGURE 8.9   ESE tool flow

cycle accurate model (CAM). The CAM is the hand-off point to standard FPGA and ASIC design automation tools. Therefore, ESE enables a structured and automated design flow from an abstract specification to an implementation, based on well defined design decisions.

The application, platform and mapping entry in ESE are simplified by an intuitive Graphical User Interface (GUI). The application is captured as a set of concurrent communicating processes. Each process has an associated C/C++ description. Channels are used to specify communication between processes. These channels provide a rich set of user level communication mechanisms, such as handshake, FIFO and asynchronous read/write.

The hardware platform is composed in the GUI from a set of processing elements (PEs), buses, and interface components called transducers. The software platform is defined by configuring the software parameters of the processing elements. These configurations include the RTOS definition, task scheduling policy and memory management. A mapping from application to platform may also be defined graphically in ESE. The C/C++ processes are mapped to PEs. Channels are mapped to buses or routes in the hardware platform.

ESE FRONT END

The goal of ESE front end is to enable fast and early design space exploration by automatically generating fast and accurate TLMs from the system specification. The details of the TLM generation process are shown in Figure 8.10. The basic idea is to automatically generate a high speed TLM that can be simulated to obtain metrics about the design; these metrics may be performance, power, reliability, security and so on. Once the metrics are obtained, the designer may

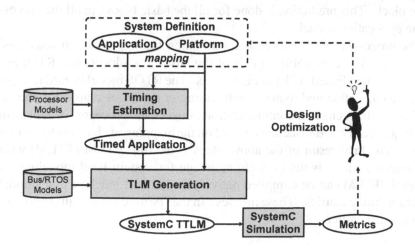

FIGURE 8.10 System level design with ESE front end

either be satisfied with them or go back to change either the application model, the platform or the mapping decisions. A practical design space exploration flow requires the capability to generate TLMs quickly. Therefore, manually coding the TLMs is not an option. TLMs must also provide reliable metrics. Perfect accuracy is desirable, but marginal error may be tolerated for a higher simulation speed.

The metric estimation supported by ESE generated TLMs is timing. Timing is annotated inside the TLM such that TLM simulation can predict timing for any input. ESE uses a retarget-able technique to automatically annotate cycle-approximate timing to the TLM. Data models of the PEs, buses and RTOS are used for timing annotation. The PE data model includes the data path and the memory hierarchy information of the PE. Therefore, it includes the number and type of architectural components and the size and configuration of caches. The bus model defines the bus transaction delays for various bus modes such as word, burst or pipelined transfer. The RTOS model includes methods for dynamic scheduling of processes and inter-process communication inside the PE.

The TLM generation occurs in two steps. The first step is the computation timing estimation where the application process code is instrumented with delays. The process code is converted into a Control Data Flow Graph (CDFG) representation. Each CDFG node represents a basic block in the application process. Based on the mapping of the process to a given PE, each basic block is statically scheduled on the PE data path. The scheduling provides the number of cycles needed to execute the basic block. The memory model of the PE is used to estimate the overhead of data and instruction cache misses. The scheduling and memory overhead delays are added to predict the delay for the basic block. This prediction is done for all the basic blocks in all the processes of the application model.

The processes, annotated with computation timing, are instantiated inside PE models. The executable models of the buses, transducers and RTOSes are instantiated and linked with the PE models. The RTOS model is used to capture resource contention and dynamic scheduling of processes mapped to the same PE. The abstract channel communication between the processes is transformed into sequence of bus transactions, based on the mapping of channels to buses and routes. The final result of the above steps is the timed TLM (TTLM) written in SystemC, which is the *de facto* language for system level modeling. The SystemC TLLM can be compiled natively on the host machine and simulated to obtain timing metrics. These metrics can then be used for design optimization as explained earlier.

ESE BACK END

After the design optimization steps are completed, a satisfactory designed is obtained at the system level. However, this design is still in the form of a TLM, which is suitable for simulation but for ready for implementation with standard EDA tools. The TLM must be transformed into the aforementioned CAM for hand-off to ASIC and FPGA implementation tools. The synthesis of the CAM from TLM is supported by the ESE back end as shown in Figure 8.11.

There are three modules in the back end, each working on different parts of the TLM. The software synthesis module produces the PE specific C/C++ code for software implementation. Naturally, the PEs in consideration for SW synthesis are embedded processors such as CPUs or DSPs. The application code is imported as is from the TLM. If an RTOS is present, the RTOS model is replaced with the actual RTOS kernel library from the database. Finally, the communication layers are generated. The communication layers implement the abstract channel based communication in the TLM using processor specific code. The synchronization with external processes is implemented using interrupt or polling. If interrupts are used, the specific interrupt handlers are generated and instantiated for each channel. If a polling option is chosen, then the HW polling flag management code is generated. The abstract data transfer of the TLM is implemented by creating an address map for the transactions and generating specific load and store transactions. Once all the code is generated, the cross-compiler for the embedded processor is used to generate the software binary.

For hardware implementation, the RTL code for the specific hardware PE must be generated. If a RTL model of the PE is already available in the IP

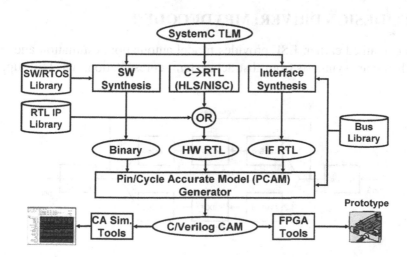

FIGURE 8.11 SW-HW synthesis with ESE back end

database, the C model in the TLM is simply replaced with this RTL model. If a RTL implementation is not available, it must be synthesized from the C model in the TLM. For this purpose, an industrial high level synthesis (HLS) tool may be used. ESE also supports generation of PE RTL model using the No Instruction Set Computer (NISC) technology [40]. The NISC technology is based on the programmable controller design of hardware PEs as explained in Section 6.1. A suitable data path template is selected based on the C profile of the process mapped to the hardware PE. Then, the NISC compiler is used to translate the C code of the process into control words to drive the data path. A Verilog RTL description of the data path and the control memory is automatically generated from the NISC tools for hardware implementation.

The final step in CAM generation is the RTL generation of the communication structure of the system. The bus protocol library is used to instantiate the bus controllers for all the buses in the system. The RTL description of all the transducers is also generated automatically based on the mapping of channels to routes in the platform. Interrupt controllers are also instantiated and configured, if needed.

The CAM produced by the ESE back end consists of C or binary code for all the software PEs in the system and RTL Verilog code for all the hardware PEs, buses and transducers. The CAM may be simulated using standard Verilog simulators available commercially. Since the Verilog code is synthesize-able, it can be input to logic synthesis tool for ASIC implementation. Alternately, ESE produces FPGA-ready description of the CAM for prototyping on FP-GAs. Therefore, ESE enables a well defined and automated path from system specification to a software/hardware implementation.

8.4.2   DESIGN DRIVER: MP3 DECODER

As explained earlier, ESE provides model automation, estimation and software/hardware synthesis from abstract system representation. The tool support

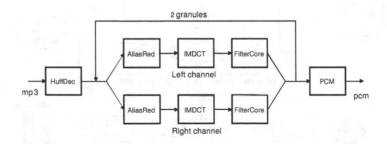

FIGURE 8.12 MP3 decoder application model

in ESE facilitates design of complex embedded systems for large applications. In order to demonstrate the efficacy of ESE, we have chosen the MP3 decoder application as a design driver. The MP3 decoder is an ideal application in many ways. It is reasonably complex, with over 13000 lines of C code, to justify a system level design approach. It is modular with well defined functions to demonstrate partitioning and hardware-software implementation. Since it typically has streaming input and output, there are real time constraints that require an application specific implementation. Finally, MP3 decoder designs are pervasive and highly relevant to mobile multimedia devices.

MP3 APPLICATION

The functional block diagram of the MP3 decoder [182] is shown in Figure 8.12. The input to the decoder is an MP3 data stream consisting of frames. Each frame of MP3 data is decoded using huffman decoding function (*HuffDec*). The frame is then split into *granules* that are sent to two channels, *left channel* and *right channel*, for stereo decoding. The two channels are data independent, so they can work on completely independent sections of the granules. Each granule section undergoes a sequence of transforms, namely alias reduction (*AliasRed*), inverse modified discrete cosine transform (*IMDCT*), and discrete cosine transform (*DCT*). Finally, the decoded granules are combined into pulse code modulated (*PCM*) frames that are ready to be sent to speaker.

In order to play the streaming MP3 file without dropped frames, the decoding rate must be at least 36 frames per second. As a result, after compensating for I/O delays, each frame must be decoded within 26.12 milliseconds (ms). Therefore, we have a real time constraint on the execution time of the decoder application. If a pure software implementation meets the required constraint, it would be an ideal implementation. Otherwise, a multi-core implementation, may be required. The decoding can be sped up by adding specialized hardware PEs for the compute intensive IMDCT and DCT functions. The data independence between the two decoding channels can also be used to parallelize the left and right channel transforms.

MP3 DESIGN FLOW

A design space exploration exercise is done with ESE to implement the MP3 decoder on a suitable platform that meets the real time constraint of 26.12 ms on the frame delay. In other words, the delay for each frame from the beginning of huffman decoding to the end of PCM output must be less than 26.12 ms. During this design space exploration, we start with a pure software implementation and incrementally move the compute intensive functions to hardware processors until the timing constraint is satisfied. The timed TLMs generated by ESE and simulated with a sample MP3 file, as input, to estimate the performance of the design and to determine if it meets the timing constraint.

The chosen underlying implementation technology is Xilinx Virtex-II FPGA [196]with a maximum clock rate of 100 MHz. For software implementation, a Xilinx Microblaze (MB) processor is used on the FPGA chip. MB interfaces with the open peripheral bus (*OPB* and an off-chip SRAM is used to store the program and data. All hardware processors are generated using the NISC tools and they interface to the double handshake bus (*DHB*). Since the OPB and DHB protocols are different, a transducer is used to interface between them. The transducer component, therefore, enables communication between MB and the hardware processors.

We start with a software implementation, in which all the MP3 functions are mapped to MB. We will refer to this mapping as *SW+0*. The 0 indicates the lack of any hardware processors. The timed TLM for $SW + 0$ was generated by ESE and the frame decoding time was estimated to be $35.66ms$. Based on this estimation, the $SW + 0$ design of the MP3 decoder does not meet the decoding time constraint.

As a next step, we decided to add a hardware processor to implement the DCT function. We will refer to the new design as *SW+1*, in which 1 refers to the DCT hardware processor. The DCT hardware is generated using the NISC tools and it uses the (*DHB*) interface protocol, as mentioned earlier. A transducer (*Tx*) was also introduced to connect OPB and DHB. The timed TLM for $SW + 1$ was generated by ESE and the frame decoding time was estimated to be $32.89ms$. Based on this estimation, the $SW + 1$ design of the MP3 decoder also does not meet the decoding time constraint. The improvement over $SW + 0$ was not too large because of the communication overhead caused by Tx.

To further improve the design performance, without much effort, we decided to use two instances of the DCT hardware processor to execute the DCT function

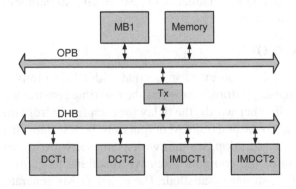

FIGURE 8.13 MP3 decoder platform SW+4

for the two decoding channels in parallel. This design is referred to as $SW+2$ because of the two hardware processors. The timed TLM for $SW + 2$ was generated by ESE and the frame decoding time was estimated to be $29.99ms$. Again, the speed up over $SW + 1$ design was only marginal. The $SW + 2$ design of the MP3 decoder also did not meet the decoding time constraint.

As a next step, we created a $SW+4$ design that included two instances each of DCT and IMDCT hardware processors. Therefore, the IMDCTs were also accelerated using specialized hardware. This platform in shown in Figure 8.13. The timed TLM for $SW + 4$ was generated by ESE and the frame decoding time was estimated to be $15.96ms$. Based on this estimation, $SW + 4$ design of the MP3 decoder met the decoding time constraint of $26.12ms$. As a result, the $SW + 4$ design was selected for implementation.

The above four platforms and mappings were created graphically in ESE and TLMs were automatically generated for evaluation of the respective designs. The TLMs were then used to synthesize software and hardware for the Microblaze soft-core processor and the Xilinx FPGA by the ESE back end. The generated software and hardware were exported to Xilinx Embedded Development Kit (EDK) for bitstream generation and programming of the FPGA. The programmed FPGA was tested with various MP3 sample inputs. In the next section, we will present various results pertaining to design of the MP3 decoder for the four platforms using ESE.

8.4.3   RESULTS

In this section we will discuss the results for system level design of the MP3 decoder with ESE. We will discuss four designs $SW+0$ to $SW+4$ as described above. The results for ESE front end demonstrate the benefits of using TLMs for early design performance estimation. The back end results demonstrate that automatic software and hardware synthesis can lead to design quality that is comparable to manual design. Automatic synthesis naturally leads to huge productivity gain in both design development and validation time. The overall design time is reduced from several months to less than a week as a result of using automatic system level design tools.

TLM ACCURACY

The MP3 design flow is simplified by the interactive graphical design decisions and automatic TLM generation. The design decisions of adding hardware processors were based on the estimation provided by the timed TLMs. Therefore, it is crucial that the TLM estimation is accurate enough for the design decisions to be made reliably. To determine if TLM estimation is accurate, let us compare the timing estimates provided by TLMs to actual board measurements for the same designs.

Figure 8.14 compares the speed and accuracy of automatically generated TLMs with traditional models. The X-axis shows the execution time of the model and the Y-axis is the relative accuracy of the timing reported by the model. The actual board design is the naturally the reference for measuring accuracy. It can be seen that the CAM provides timing estimation that is identical to the board measurements. Since the CAM is cycle accurate, this is to be expected. However, the simulation time of the CAM is in the order of 15 to 18 hours for each MP3 sample frame. This is inordinately long for any reasonable design space exploration.

Typically, designers use instruction set simulation model (*ISM*) of a processor to speed up simulation. An ISM models the processor micro-architecture in a high level language such as C/C++. The binary of the software is loaded into the ISM memory. During simulation, the ISM interprets the instruction stream and updates the processor state. The hardware peripherals may be modeled in RTL using VHDL or Verilog. The high level processor model is instantiated as a module in Verilog. The ISM is typically faster than the CAM because it does not model the processor at the cycle-accurate level. However, the performance estimation accuracy of the ISM may vary based on the quality of the processor model. In the case of the MP3 designs, the accuracy of the ISM varied from 50% to 80% compared to board measurements. The unpredictable accuracy of

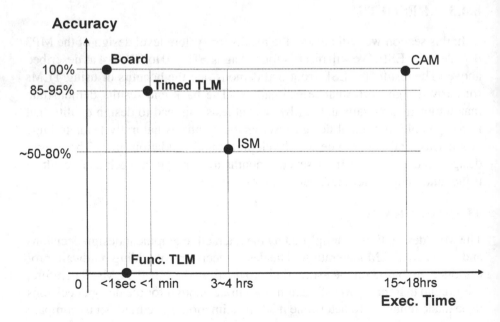

FIGURE 8.14. Execution speed and accuracy trade-offs for embedded system models

ISMs makes them unsuitable for early design space exploration. Furthermore, although the simulation speed of ISMs was about 5 times faster than the CAMs, it was still in the order of few hours.

The TLMs generated automatically by the ESE front end were two orders of magnitude faster than the ISM or the CAM. The timed TLMs were generated for all the design in under a minute and simulated under a minute as well. In contrast with the ISM, the timed TLMs were consistently accurate for all the platforms. A marginal error of under 15% was found in the TLM performance estimation. Therefore, designers can use timed TLMs for early estimation with a high degree of confidence.

In Figure 8.14, we distinguish between timed and untimed (or functional) TLMs. While the timed TLMs are used for performance estimation, the high simulation speed of functional TLMs makes them ideal for software development. It must be noted that functional TLMs may be generated even for a partial or test application. The process code may be developed using the functional TLM as a virtual platform. The results therefore demonstrate the efficacy and suitability of TLMs for early application development and reliable performance estimation.

DESIGN QUALITY

One of the primary concerns of automatic synthesis methods is the quality of design. Various metrics for design quality may be used. Some of the most common metrics are performance, silicon footprint and power dissipation. Generally speaking, it is difficult to evaluate the efficiency of a synthesis method by comparing its output to a manual design. The manual design is highly sen-

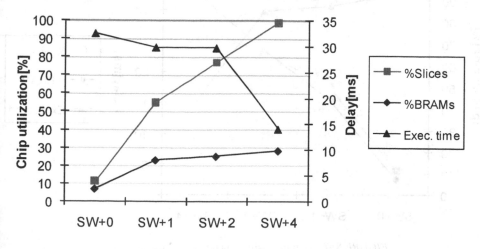

FIGURE 8.15 MP3 manual design quality

the designer. Nonetheless, a comparison of synthesized designs with an expert manual designer may give us better insight into the industrial viability of the synthesis tool.

To evaluate the quality of implementation produced by ESE back end, an expert designer created the software/hardware implementations of the four MP3 decoder designs described earlier. The hardware PEs were designed in RTL and the software was implemented directly on the FPGA using the Xilinx EDK tools. Figure 8.15 shows the performance and area of the manual designs.

In order to evaluate the performance of the designs, a sample MP3 file was loaded on the on-board memory and used as input. The average decoding time for each frame is shown in milliseconds. The pure software design is too slow to meet the 26.12 millisecond decoding time constraint. As predicted by TLM simulation, only the $SW + 4$ implementation was able to meet the specified real time constraint. The design area is indicated by the percentage of block RAMs (BRAMs) and FPGA slices used by the implementation. The hardware PEs, namely the DCT and the IMDCT, had a hardwired controller implementation, which justifies the high number of slices used by the $SW + 4$ implementation.

Figure 8.16 shows the performance and area of designs generated automatically from ESE. Compared to the corresponding manual designs, the performance of the generated designs was almost identical. In this case too, only the $SW + 4$ design was able to meet the real time constraints imposed by the application. The area of the generated designs was different compared the manual designs. Notably, fewer slices were used in the generated design but

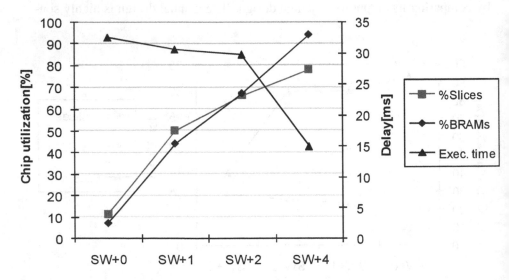

FIGURE 8.16 Automatically generated MP3 design quality

template was used for the hardware PEs in automatically generated designs. In contrast to the manual designed hardware PEs, NISC uses control words in memory to drive the data path. Therefore, NISC implementations are generally memory intensive. However, all the design could still fit on the target Virtex-II FPGA. The total number of FPGA resources used by automatically generated designs was comparable to the manual designs. Therefore, we can conclude that automatically generated designs are comparable to manual designs in terms of quality metrics of performance and area. This is a significant argument in favor of using automatic system level design tools.

PRODUCTIVITY GAINS

The single most important factor that drives the rise in design abstraction level is productivity gain. Typically, designs descriptions at higher abstraction levels are more compact, understandable and easily modified. Therefore, greater optimization opportunities are available at higher abstraction level. The two key productivity metrics we consider here are the design development time and validation time. Development time directly translates to design cost and time to market. Naturally, reducing the development time is always desirable. Similarly, design validation time directly impacts quality of design which is an important factor is product success.

Figure 8.17 illustrates the productivity gain in development time as a result of using ESE. Traditional design practice starts with RTL and embedded SW coding for selected platforms. The reference C specification model is used for developing test bench to verify the cycle accurate models. For MP3 platforms

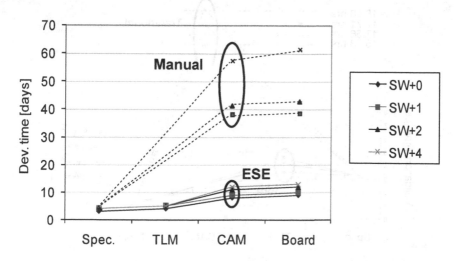

FIGURE 8.17 Development productivity gains from model automation

with HW components, the RTL development time was in the order of months. As a result, board prototypes for these designs took between 40 to 60 days. ESE drastically cuts prototype development time by automatically generating TLM and RTL models. With ESE, the final board prototypes for MP3 designs were available in less than a week after the specification model was finalized. Consequently, ESE results in significant savings in design cost and shorter development cycles.

Figure 8.18 illustrates the productivity gain resulting from a TLM based design methodology supported by ESE. As a consequence of traditional cycle accurate modeling, designers must make design optimizations and changes on RTL and low level SW code. Each change needs to be verified using time consuming cycle accurate simulations. Each CAM simulation of the MP3 designs took 15 to 18 hours for MP3 designs. This is a significant component of design time. Although at speed on-board verification is faster than even reference application C model simulation, bugs found in on-board testing are difficult to trace back to the CAM.

TLMs remove the burden of cycle accurate simulations by moving the design abstraction to a higher level. ESE generated TLMs execute at the same speed as reference C simulation. Design changes can made at the transaction level and can hence be verified and debugged using the automatically generated high speed TLMs. TLMs are easier to debug and maintain because their code size is at least an order of magnitude less than corresponding CAM code size.

Automatic CAM generation from TLM is also less likely to introduce bugs in the design compared to manual CAM optimizations. This has been true in

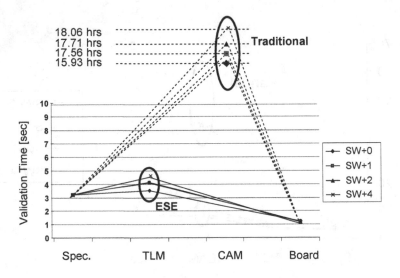

FIGURE 8.18 Validation productivity gain from using TLM vs. CAM

the past when the modeling abstraction moved from gate level to RTL with the use of logic synthesis tools. Therefore, ESE reduces validation time from an order of several hours or even days to a few seconds. As a results, designers can use ESE to make platform and application optimizations at a higher level, automatically generate TLMs and verify the optimizations in a few seconds.

8.5 SUMMARY

We discussed several academic and commercial tools for various aspects of embedded system design. These range from system level modeling and simulation to automatic synthesis of software and hardware. We also presented a case study for design of MP3 decoder on a heterogeneous platform. The results show that the design methods presented in this book can work for practical embedded system design. The automatic design tools provide fast and accurate models, design quality comparable to manual and huge productivity gains. These results point to the significant advantages and benefits of using embedded system design methods described in this book.

References

[1] Samar Abdi and Daniel Gajski. Functional validation of system level model transformations. *International Journal of Parallel Programming*, 34(1):29–59, February 2006.

[2] Samar Abdi, Dongwan Shin, and Daniel Gajski. Automatic communication refinement for system level design. In *Design Automation Conference*, pages 300–305, 2003.

[3] Accellera. *RTL Semantics: Draft Specification, Version 0.8*. Working Group of the Architectural Language Committee, February 2001.

[4] Advanced RISC Machines Ltd. (ARM). *AMBA Specification (Revision 2.0)*, 1999. ARM IHI 0011A.

[5] Perry Alexander. *System Level Design with Rosetta*. Morgan Kaufmann, 2006.

[6] Charles André. Representation and analysis of reactive behaviors: A synchronous approach. In *Computational Engineering in System Applications (CESA)*, Lille, France, July 1996.

[7] Peter J. Ashenden. *The Designer's Guide to VHDL*. Morgan Kaufmann, December 1995.

[8] Motor Industry Research Association. *MISRA-C 2004: Guidelines for the Use of the C Language in Critical Systems*. 2004.

[9] AUTOSAR Partnership. Autosar: Automotive open system architecture. http://www.autosar.org/.

[10] M. Balakrishnan and P. Marwedel. Integrated scheduling and binding: A synthesis approach for design space exploration. In *Design Automation Conference*, pages 68–74, Las Vegas, NV, June 1989.

[11] Felice Balarin, Massimiliano Chiodo, Paolo Giusto, Harry Hsieh, Attila Jurecska, Luciano Lavagno, Claudio Passerone, Alberto Sangiovanni-Vincentelli, Ellen Sentovich, Kei Suzuki, and Bassam Tabbara. *Hardware-Software Co-Design of Embedded Systems: The POLIS Approach*. Kluwer Academic Publishers, 1997.

[12] Felice Balarin, Harry Hsieh, Luciano Lavagno, Claudio Passerone, Alessandro Pinto, Alberto Sangiovanni-Vincentelli, Yosinori Watanabe, and Guang Yang. Metropolis: A

design environment for heterogeneous systems. In Wayne Wolf and Ahmed Jerraya, editors, *Multiprocessor Systems-on-Chips*. Morgan Kaufmann, 2004.

[13] M. Barbacci. Instruction set processor specification (isps): The notation and its application. *IEEE Transactions on Computers*, C-30(1):24–40, January 1981.

[14] M. R. Barbacci. A comparison of register transfer level languages for describing computers and other digital systems. *IEEE Transactions on Computers*, C-24(2), February 1975.

[15] M. R. Barbacci. Instruction set processor specifications for simulation, evaluation, and synthesis. In *Design Automation Conference*, pages 64–72, San Diego, CA, United States, 1979.

[16] C. G. Bell and A. Newell. *Computer Structures: Readings and Examples*. McGraw-Hill, 1971.

[17] Rudy Belliardi, Ben Brosgol, Peter Dibble, David Holmes, and Andy Wellings. *The Real-Time Specification for Java*, 2006.

[18] Albert Beneviste, Paul Caspi, Stephen A. Edwards, Nicolas Halbwachs, Paul Le Guernic, and Robdert de Simone. The synchronous languages twelve years later. *Proceedings of the IEEE*, 91(1):64–83, January 2003.

[19] L. Bening and H. Foster. *Principles of Verifiable RTL Design*. Kluwer Academic Publishers, 2000.

[20] Luca Benini, Davide Bertozzi, Alessandro Bogliolo, Francsco Menichelli, and Mauro Olivieri. MPARM: Exploring the multi-processor SoC design space with SystemC. *Journal of VLSI Signal Processing*, 41(2):169–182, September 2005.

[21] Gerard Berry. The foundations of Esterel. In Gordon Plotkin, Colin Stirling, and Mads Tofte, editors, *Proof, Language, and Interaction: Essays in Honor of Robin Milner*. MIT Press, 2000.

[22] Greet Bilsen, Marc Engels, Rudy Lauwereins, and Jean Peperstraete. Cyclo-static dataflow. *IEEE Transactions on Signal Processing*, 44(2):397–408, February 1996.

[23] Grady Booch, Ivar Jacobson, and James Rumbaugh. *Unified Modeling Language (UML) Specification, Version 1.5*. Object Management Group (OMG), March 2003.

[24] Aimen Bouchhima, Iuliana Bacivarov, Wassim Youssef, Marius Bonaciu, and Ahmed A. Jerraya. Using abstract CPU subsystem simulation model for high level HW/SW architecture exploration. In *Asia and South Pacific Design Automation Conference (ASP-DAC)*, Shanghai, China, January 2005.

[25] R. K. Brayton, R. Camposano, G. De Micheli, R.H.J.M. Otten, and J. Van Eijndhoven. The yorktown silicon compiler system. In Daniel D. Gajski, editor, *Silicon Compilation*. Addison-Wesley, 1988.

[26] Forrest D. Brewer and Daniel D. Gajski. An expert-system paradigm for design. In *Design Automation Conference*, pages 203–509, Las Vegas, NV, June 1986.

[27] R.E. Bryant. Graph-based algorithms for boolean function manipulation. *IEEE Transactions on Computer*, C-35(8):677–691, August 1986.

[28] Joseph Buck, Soonhoi Ha, Edward A. Lee, and David G. Messerschmitt. Ptolemy: A framework for simulating and prototyping heterogeneous systems. *International Journal of Computer Simulation, Special Issue on Simulation Software Development*, 4:155–182, April 1994.

[29] David R. Butenhof. *Programming with POSIX Threads*. Addison-Wesley, 1997.

[30] Giorgio C. Buttazzo. *Hard Real-Time Computing Systems*. Kluwer Academic Publishers, 1999.

[31] Lukai Cai and Daniel Gajski. Transaction level modeling: An overview. In *International Symposium on Hardware/Software Codesign and System Synthesis (CODES+ISSS)*, Newport Beach, CA, USA, October 2003.

[32] Lukai Cai, Andreas Gerstlauer, and Daniel Gajski. Retargetable profiling for rapid, early system-level design space exploration. In *Design Automation Conference*, San Diego, CA, USA, June 2004.

[33] Jean-Paul Calvez. *Embedded Real-Time Systems: A Specification and Design Methodology*. John Wiley and Sons, 1993.

[34] Raul Camposano. Path-based scheduling for synthesis. *IEEE Transactions on Computer-Aided Design of Integrated Circuits and Systems*, 10(1):85–93, January 1991.

[35] Raul Camposano and Wayne Wolf (editors). *High-Level VLSI Synthesis*. Kluwer Academic Publishers, 1991.

[36] Raul Camposano and Wolfgang Rosenstiel. A design environment for the synthesis of integrated circuits. In *EUROMICRO Symposium on Microprocessing and Microprogramming*, Brussels, Belgium, September 1985.

[37] Carbon Design Systems. Carbon SoC Designer. http://www.carbondesignsystems.com/.

[38] Celoxica Ltd. *Handel-C Language Reference Manual*, 2005.

[39] Center for Embedded Computer Systems (CECS). Embedded System Environment, Center for Embedded Computer Systems, University of California, Irvine. http://www.cecs.uci.edu/~ese, 2008.

[40] Center for Embedded Computer Systems (CECS). NISC Technology. http://www.cecs.uci.edu/~nisc/, 2008.

[41] D. Chen, J. Cong, Y. Fan, G. Han, W. Jiang, and Z. Zhang. xpilot: A platform-based behavioral synthesis system. In *SRC Techcon Conference*, October 2005.

[42] E. M. Clarke, O. Grumberg, and D. A. Peled. *Model Checking*. MIT Press, January 2000.

[43] CoFluent Design. CoFluent Studio. http://www.cofluentdesign.com/.

[44] Lockheed Martin Corporation. *JSF Air Vehicle C++ Coding Standards for the System Development and Demonstration Program*, 2005.

[45] P. Coussy and A. Morawiec, editors. *High-Level Synthesis: from Algorithm to Digital Circuit*. Springer, 2008.

[46] CoWare. http://www.coware.com/.

[47] G. de Jong. A uml-based design methodology for real-time and embedded systems. In *IEEE International Conference Design and Test in Europe (DATE)*, pages 776–779, Paris, France, March 2002.

[48] S. Devadas, H.K. T. Ma, and A. R. Newton. On the verification of sequential machines at different levels of abstraction. In *Design Automation Conference*, pages 271–276, Miami Beach, FL, USA, June 1987.

[49] Srinivas Devadas and A. Richard Newton. Algorithm for allocation in data path synthesis. *IEEE Transactions on Computer-Aided Design of Integrated Circuits and Systems*, 8(7):768–781, July 1989.

[50] S. W. Director, A. C. Parker, D. P. Siewiorek, and D. E. Thomas. A design methodology and computer design aids for digital vlsi systems. *IEEE Transactions on Circuits and Systems*, 28(7):634–645, July 1981.

[51] Rainer Dömer, Andreas Gerstlauer, and Daniel Gajski. *SpecC Language Reference Manual, Version 2.0*. SpecC Technology Open Consortium (STOC), 2002.

[52] Rainer Dömer, Andreas Gerstlauer, Junyu Peng, Dongwan Shin, Lukai Cai, Haobo Yu, Samar Abdi, and Daniel Gajski. System-on-Chip Environment: A SpecC-based Framework for Heterogeneous MPSoC Design. *EURASIP Journal on Embedded Systems (JES)*, 2008(647953):13, 2008.

[53] dSPACE (Digital Signal Processing And Control Engineering). TargetLink. http://www.dspace.com/.

[54] Bruce Eckel. *Thinking in Java*. Prentice-Hall, Upper Saddle River, N.J., 2003.

[55] Stephen A. Edwards. *Languages for Digital Embedded Systems*. Kluwer Academic Publishers, 2000.

[56] J. P. Elliot. *Understanding Behavioral Synthesis: A Practical Guide to High-Level Design*. Kluwer Academic Publishers, 1999.

[57] Esterel Technologies. Scade suite. http://www.esterel-technologies.com/.

[58] Forte Design Systems. Cynthesizer. http://www.forteds.com/, 2008.

[59] Eclipse Foundation. Eclipse. http://www.eclipse.org/.

[60] D. Gajski and R. Kuhn. New vlsi tools. *Computer Magazine*, pages 11–14, December 1983.

[61] Daniel Gajski, Nikil Dutt, Allan Wu, and Steve Lin. *High-Level Synthesis: Introduction to Chip and System Design*. Kluwer Academic Publishers, 1992.

[62] Daniel D. Gajski. *Principles of Digital Design*. Prentice-Hall, September 1996.

[63] Daniel D. Gajski, Frank Vahid, Sanjiv Narayan, and Jie Gong. *Specification and Design of Embedded Systems*. Prentice-Hall, July 1994.

[64] Daniel D. Gajski, Frank Vahid, Sanjiv Narayan, and Jie Gong. SpecSyn: An environment supporting the specify-explore-refine paradigm for hardware/software system design. *IEEE Transactions on Very Large Scale Integrated Systems (TVLSI)*, 6(1):84–100, March 1998.

[65] Daniel D. Gajski, Jianwen Zhu, Rainer Doemer, Andreas Gerstlauer, and Shuqing Zhao. *SpecC: Specification Language and Methodology*. Kluwer Academic Publishers, March 2000.

[66] Lovic Gauthier, Sungjoo Yo, and Ahmed A. Jerraya. Automatic Generation and Targeting of Application-Specific Operating Systems and Embedded Systems Software. *IEEE Transactions on Computer-Aided Design of Integrated Circuits and Systems*, 20(11), November 2001.

[67] Geensys. Autosar builder. http://www.geensys.com/.

[68] Marc Geilen and Twan Basten. Requirements on the execution of Kahn process networks. In *European Symposium on Programming (ESOP)*, pages 319–334, Warsaw, Poland, April 2003.

[69] Gentleware. Poseidon for uml. http://www.gentleware.com/.

[70] Patrice Gerin, Sungjoo Yoo, Gabriela Nicolescu, and Ahmed A. Jerraya. Scalable and flexible cosimulation of SoC designs with heterogeneous multiprocessor target architectures. In *Asia and South Pacific Design Automation Conference (ASP-DAC)*, Yokohama, Japan, January 2001.

[71] Andreas Gerstlauer. *Modeling Flow for Automated System Design and Exploration*. PhD thesis, Information and Computer Science, University of California, Irvine, May 2004.

[72] Andreas Gerstlauer, Rainer Dömer, Junyu Peng, and Daniel D. Gajski. *System Design: A Practical Guide with SpecC*. Kluwer Academic Publishers, 2001.

[73] Andreas Gerstlauer, Junyu Peng, Dongwan Shin, Daniel Gajski, Atsushi Nakamura, Dai Araki, and Yuuji Nishihara. Specify-Explore-Refine (SER): From specification to implementation. In *Proceedings of the Design Automation Conference (DAC)*, pages 586–591, Anaheim, CA, USA, June 2008.

[74] Andreas Gerstlauer, Dongwan Shin, Junyu Peng, Rainer Doemer, and Daniel Gajski. Automatic, layer-based generation of system-on-chip bus communication models. *IEEE Transactions on Computer-Aided Design of Integrated Circuits and Systems*, 26(9):1676–1687, September 2007.

[75] Andreas Gerstlauer, Haobo Yu, and Daniel D. Gajski. RTOS modeling for system level design. In Ahmed A. Jerraya, Sungjoo Yu, Norbert Wehn, and Diedrik Verkest, editors, *Embedded Software for SoC*. Springer, September 2003.

[76] Andreas Gerstlauer, Shuqing Zhao, Daniel Gajski, and Arkady Horak. Specc system-level design methodology applied to the design of a gsm vocoder. In *SASIMI*, 2000.

[77] Frank Ghenassia, editor. *Transaction-Level Modeling with SystemC: TLM Concepts and Applications for Embedded Systems.* Springer, November 2005.

[78] Gordon. Specification and verification of hardware, October 1992.

[79] James Gosling, Bill Joy, Guy L. Steele Jr., and Gilad Bracha. *The Java Language Specification.* Addison-Wesley, third edition, 2005.

[80] William Gropp, Ewing Lusk, and Anthony Skjellum. *Using MPI: Portable Parallel Programming with the Message Passing Interface.* MIT Press, second edition, 1999.

[81] Torsten Grötker, Stan Liao, Grant Martin, and Stuart Swan. *System Design with SystemC.* Springer, 2002.

[82] Yuri Gurevich. Evolving algebras 1993: Lipari guide. In Egon Börger, editor, *Specification and Validation Methods.* Oxford University Press, 1995.

[83] Soonhoi Ha, Sungchan Kim, Choonseung Lee, Youngmin Yi, Seongnam Kwon, and Young-Pyo Joo. PeaCE: A hardware-software codesign environment of multimedia embedded systems. *ACM Transactions on Design Automation of Electronic Systems (TODAES)*, 12(3):1–25, 2007.

[84] L. Hafer and A. C. Parker. Register transfer level automatic digital design: The allocation process. In *Design Automation Conference*, Las Vegas, NV, United States, June 1978.

[85] Nicolas Halbwachs, Paul Caspi, Pascal Raymond, and Daniel Pilaud. The synchronous dataflow programming language Lustre. *Proceedings of the IEEE*, 79(9):1305–1320, September 1991.

[86] David Harel. Statecharts: A visual formalism for complex systems. *Science of Computer Programming*, 8(3):231–274, June 1987.

[87] David Harel and Amnon Naamad. The STATEMATE semantics of Statecharts. *ACM Transactions on Software Engineering and Methodology (TOSEM)*, 5(4):293–333, October 1996.

[88] Graham Hellestrand. The engineering of supersystems. *IEEE Computer*, 38(1):103–105, January 2005.

[89] F. Herrera, H. Posadas, P. Sanchez, and E. Villar. Systematic Embedded Software Generation from SystemC. In *Proceedings of the Design Automation and Test Conference in Europe*, Munich, Germany, March 2003.

[90] C. A. R. Hoare. *Communicating Sequential Processes.* Prentice-Hall, 1985.

[91] Andreas Hoffmann, Heinrich Meyr, and Rainer Leupers. *Architecture Exploration for Embedded Processors with LISA.* Kluwer Academic Publishers, 2003.

[92] Matthias Homann. *OSEK: Betriebssystem-Standard für Automotive und Embedded Systems.* mitp-Verlag, 2 edition, 2005.

[93] Yonghyun Hwang, Samar Abdi, and Daniel Gajski. Cycle-approximate retargetable performance estimation at the transaction level. In *IEEE International Conference Design and Test in Europe (DATE)*, pages 3–8, Munich, Germany, March 2008.

[94] IBM. Telelogic rhapsody. http://www.ibm.com/.

[95] MathWorks Inc. *MATLAB and Simulink Student Edition*. Pearson, 2008.

[96] National Instruments Inc. and Robert H. Bishop. *LabVIEW Student Edition*. Prentice-Hall, 2007.

[97] Tensilica Inc. Xtensa xplorer design environment. http://tensilica.com/.

[98] International Organization for Standardization. *Reference Model of Open System Interconnection (OSI)*, second edition, 1994. ISO/IEC 7498 Standard.

[99] International Technology Roadmap for Semiconductors (ITRS). ITRS Home. http://www.itrs.net/, 2008.

[100] R. S. Janka. *Specification and Design Methodology for Real-Time Embedded Systems*. Kluwer Academic Publishers, 2004.

[101] Axel Jantsch. *Modeling Embedded Systems and SoCs: Concurrency and Time in Models of Computation*. Morgan Kaufmann, 2004.

[102] A. A. Jerraya, H. Ding, P. Kission, and M. Rahmouni. *Behavioral Synthesis and Component Reuse with VHDL*. Kluwer Academic Publishers, 1997.

[103] Ahmed A. Jerraya. Long term trends for embedded system design. In *EUROMICRO Symposium on Microprocessing and Microprogramming*, pages 20–26, Rennes, France, September 2004.

[104] Gilles Kahn. The semantics of a simple language for parallel programming. In *Information Processing*, pages 471–475, Stockholm, Sweden, August 1974.

[105] Joachim Keinert, Martin Streubühr, Thomas Schlichter, Joachim Falk, Jens Gladigau, Christian Haubelt, Jürgen Teich, and Mike Meredith. SystemCoDesigner - an automatic ESL synthesis approach by design space exploration and behavioral synthesis for streaming applications. *ACM Transactions on Design Automation of Electronic Systems (TODAES)*, 14(1):1–23, 2009.

[106] Brian Kernighan and Dennis Ritchie. *The C programming language*. Prentice-Hall, Englewood Cliffs, NJ, 1988.

[107] Kurt Keutzer, Sharad Malik, Richard A. Newton, Jan M. Rabaey, and Alberto Sangiovanni-Vincentelli. System-level design: Orthogonalization of concerns and platform-based design. *IEEE Transactions on Computer-Aided Design of Integrated Circuits and Systems*, 19(12):1523–1543, December 2000.

[108] A. A. Khan, Carolyn McCreary, and M. S. Jones. A comparison of multiprocessor scheduling heuristics. In *International Conference on Parallel Processing*, pages 243–250, 1994.

[109] Wolfgang Klingauf, Robert Günzel, Oliver Bringmann, Pavel Parfuntesu, and Mark Burton. GreenBus: A generic interconnect fabric for transaction-level modeling. In *Design Automation Conference*, San Francisco, CA, USA, July 2006.

[110] D. W. Knapp. *Behavioral Synthesis: Digital System Design Using the Synopsys Behavioral Compiler*. Prentice-Hall, 1996.

[111] Hermann Kopetz. *Real-Time Systems: Design Principles for Distributed Applications.* Kluwer Academic Publishers, 1997.

[112] Matthias Krause, Oliver Bringmann, and Wolfgang Rosenstiel. Target software generation: An approach for automatic mapping of SystemC specifications onto real-time operating systems. 10(4):229–251, December 2005.

[113] T. Kropf. *Introduction to Formal Hardware Verification.* Springer, 1999.

[114] D. Ku and G. De Micheli. *High Level Synthesis of ASICs under Timing and Synchronization Constraints.* Kluwer Academic Publishers, 1992.

[115] Seongnam Kwon, Yongjoo Kim, Woo-Chul Jeun, Soonhoi Ha, and Yunheung Paek. A retargetable parallel programming framework for MPSoC. *ACM Transactions on Design Automation of Electronic Systems (TODAES)*, 13(3), 2008.

[116] Luciano Lavagno, Grant Martin, and Bran Selic, editors. *UML for Real: Design of Embedded Real-Time Systems.* Kluwer Academic Publishers, Norwell, MA, USA, 2003.

[117] Luciano Lavagno, Alberto Sangiovanni-Vincentelli, and Ellen Sentovich. Models of computation for embedded system design. In Ahmed Jerraya and Jean Mermet, editors, *System-Level Synthesis.* Kluwer Academic Publishers, 1999.

[118] Edward A. Lee. Consistency in dataflow graphs. *IEEE Transactions on Parallel and Distributed Systems*, 2(2):223–235, April 1991.

[119] Edward A. Lee. The problem with threads. *IEEE Computer*, 39(5):33–42, May 2006.

[120] Edward A. Lee and David G. Messerschmitt. Synchronous data flow. *Proceedings of the IEEE*, 75(9):1235–1245, September 1987.

[121] INMOS Limited. *Occam 2 Reference Manual.* Prentice-Hall, 1988.

[122] Joe S. Lis and Daniel D. Gajski. Synthesis from vhdl. In *IEEE International Conference on Computer Design*, 1988.

[123] Lucky Lo Chi Yu Lo and Samar Abdi. Automatic systemc tlm generation for custom communication platforms. In *International Conference on Computer Design*, pages 41–46, 2007.

[124] H. De Man, J. Rabaey, P. Six, and L. Claesen. Cathedral-II: A Silicon Compiler for Digital Signal Processing. *IEEE Design and Test of Computers*, 3(6):13–25, November 1986.

[125] Florence Maraninch. The Argos language: Graphical representation of automata and description of reactive systems. In *International Conference on Visual Languages*, Kobe, Japan, October 1991.

[126] Grant Martin and Wolfgang Müller, editors. *UML for SOC Design.* Springer, 2005.

[127] Peter Marwedel. The MIMOLA design system: Detailed description of the software system. In *Design Automation Conference*, pages 59–63, San Diego, CA, United States, June 1979.

[128] Peter Marwedel. A new synthesis algorithm for mimola software system. In *Design Automation Conference*, pages 131–137, Las Vegas, NV, June 1986.

[129] Peter Marwedel. *Embedded Systems Design*. Kluwer Academic Publishers, 2003.

[130] Peter Marwedel. *Embedded System Design*. Springer, 2006.

[131] MathWorks Inc. Real-Time Workshop. http://www.mathworks.com/.

[132] MathWorks Inc. Simulink - Simulation and Model-Based Design. http://www.mathworks.com/.

[133] M. C. McFarland. The value trace: A database for automated digital design. Master's thesis, Carnegie-Mellon University, December 1978.

[134] M. C. McFarland. Using bottom-up design technique in the synthesis of digital hardware from abstract behavioral descriptions. In *Design Automation Conference*, Las Vegas, NV, June 1986.

[135] M.C. McFarland. Formal verification of sequential hardware: A tutorial. *IEEE Transactions on Computer-Aided Design of Integrated Circuits and Systems*, 12(5):633–653, May 1993.

[136] K.L. McMillan. *Symbolic Model Checking: An approach to the State Explosion Problem*. Kluwer Academic Publishers, 1993.

[137] Mentor Graphics. The EDA Technology Leader - Mentor Graphics. http://www.mentor.com/, 2008.

[138] P. Michel, U. Lauther, and P. Duzy, editors. *Synthesis Approach to Digital System Design*. Kluwer Academic Publishers, 1992.

[139] Giovanni De Micheli. *Synthesis and Optimization of Digital Circuits*. McGraw-Hill, 1994.

[140] Microelectronic Embedded Systems Laboratory. SPARK: High-Level Synthesis using Parallelizing Compiler Techniques. http://mesl.ucsd.edu/spark/, 2008.

[141] Robin Milner. *A Calculus of Communicating Systems*. Springer, 1980.

[142] Tadao Murata. Petri nets: Properties, analysis and applications. *Proceedings of the IEEE*, 77(4):541–580, April 1989.

[143] Andre Nacul and Tony Givargis. Synthesis of Time-Constrained Multitasking Embedded Software. volume 11, pages 822–847, October 2006.

[144] NEC System Technologies Ltd. CyberWorkBench - System LSI Design Environment to implement All-in-C Concept. http://www.necst.co.jp/, 2008.

[145] H. Nikolov, M. Thompson, T. Stefanov, A. D. Pimentel, S. Polstra, R. Bose, C. Zissulescu, and E. F. Deprettere. Daedalus: Toward composable multimedia MP-SoC design. In *Proc. of the ACM/IEEE Int. Design Automation Conference (DAC '08)*, pages 574–579, June 2008.

[146] Achim Nohl, Gunnar Braun, Oliver Schliebusch, Rainer Leupers, Heinrich Meyr, and Andreas Hoffmann. A universal technique for fast and flexible instruction-set architecture simulation. In *Design Automation Conference*, New Orleans, LA, USA, June 2002.

[147] Object Management Group (OMG). Unified modeling language (UML). http://www. uml.org/.

[148] Object Management Group (OMG). *Common Object Request Broker Architecture: Core Specification, Version 3.0.3*, 2004.

[149] Object Management Group (OMG). *OMG Systems Modeling Language (OMG SysML), Version 1.1*, 2008.

[150] Open SystemC Initiative (OSCI). http://www.systemc.org/, 2008.

[151] Alex Orailoglu and Daniel D. Gajski. Flow graph representation. In *Design Automation Conference*, pages 503–509, Las Vegas, NV, June 1986.

[152] Barry M. Pangrle and Daniel D. Gajski. Design tools for intelligent silicon compilation. *IEEE Transactions on Computer-Aided Design of Integrated Circuits and Systems*, 6(6):1098–1112, November 1987.

[153] Barry M. Pangrle and Daniel D. Gajski. Slicer: A state synthesizer for intelligent silicon compilation. In *IEEE International Conference on Computer Design*, Rye Brook, NY, October 1987.

[154] Thomas M. Parks. *Bounded Scheduling of Process Networks*. PhD thesis, Electrical Engineering and Computer Science, University of California, Berkeley, December 1995.

[155] P. Paulin and J. P. Knight. Algorithms for high-level synthesis. *IEEE Computer*, 6(6):18–31, November 1989.

[156] P. G. Paulin and J. P. Knight. Force-directed scheduling for behavioral synthesis of asic's. *IEEE Transactions on Computer-Aided Design of Integrated Circuits and Systems*, 8(6):661–679, June 1989.

[157] Philippe Coussy and Dominique Heller. GAUT WEB SITE. http://web.univ-ubs.fr/ gaut, 2008.

[158] Gordon D. Plotkin. A structural approach to operational semantics. *Journal of Logic and Algebraic Programming*, 60(61):17–139, July 2004.

[159] Chris Porthouse. Jazelle for execution environments. http://www.arm.com/pdfs/ JazelleRCTWhitePaper_final1-0_.pdf, May 2005.

[160] J. Rabaey, H. De Man, J. Vanhoof, G. Goossens, and F. Catthoor. Cathedral-II: A Synthesis System for Multiprocessor DSP Systems. In Daniel D. Gajski, editor, *Silicon Compilation*. Addison-Wesley, 1988.

[161] Mehrdad Reshadi and Daniel Gajski. A cycle-accurate compilation algorithm for custom pipelined datapaths. In *International Symposium on Hardware/Software Codesign and System Synthesis (CODES+ISSS)*, 2005.

[162] Sebastian Ritz, Matthias Pankert, Vojin Zivojnvic, and Heinrich Meyr. High-Level Software Synthesis for the Design of Communication Systems. *IEEE Journal on Selected Areas in Communications*, April 1993.

[163] Stewart Robinson. *Simulation: The Practice of Model Development and Use.* John Wiley and Sons, March 2004.

[164] Alberto Sangiovanni-Vincentelli. Quo Vadis SLD: Reasoning about the Trends and Challenges of System Level Design. *Proceedings of the IEEE*, 95(3):467–506, March 2007.

[165] Alberto Sangiovanni-Vincentelli and Grant Martin. The platform-based design and software design methodology for embedded systems. *IEEE Design and Test of Computers*, 18(6):23–33, November 2001.

[166] Gunar Schirner, Andreas Gerstlauer, and Rainer Doemer. Abstract, multifaceted modeling of embedded processors for system level design. In *Asia and South Pacific Design Automation Conference (ASP-DAC)*, Yokohama, Japan, January 2007.

[167] Dana Scott and Christopher Strachey. Toward a mathematical semantics for computer languages. Technical Report PRG-6, Oxford Programming Research Group, 1971.

[168] D. P. Siewiorek and M. R. Barbacci. The cmu rt-cad system: An innovative approach to computer aided design. In *AFIPS National Computer Conference*, pages 643–655, New York, NY, United States, June 1976.

[169] Artisan Software. Artisan studio. http://www.artisansoftwaretools.com/.

[170] Space Codesign Systems. http://www.spacecodesign.com/.

[171] SpecC Technology Open Consortium Office. SpecC Technology Open Consortium. http://www.specc.gr.jp/, 2008.

[172] The SPIRIT Consortium. *IP-XACT, Release 1.4*, March 2008.

[173] Bjarne Stroustrup. *The C++ Programming Language.* Addison-Wesley, Reading, MA, 1997.

[174] Stuart Sutherland, Simon Davidmann, and Peter Flake. *SystemVerilog For Design: A Guide to Using SystemVerilog for Hardware Design and Modeling.* Springer, June 2003.

[175] Spark Systems. Enterprise architect. http://www.sparxsystems.com.au/.

[176] D. E. Thomas, E. M. Dirkes, R. A. Walker, J. V. Rajan, J. A. Nestor, and R. L. Blackburn. The system architect's workbench. In *Design Automation Conference*, pages 337–343, Anaheim, CA, June 1988.

[177] D. E. Thomas, C. Y. Hitchcock, T. J. Kowalski, J. V. Rajan, and R. A. Walker. Method of automatic data path synthesis. *IEEE Computer*, 16(12):59–70, December 1983.

[178] D. E. Thomas, E. D. Lagnese, R. A. Walker, J. A. Nestor, J.V Rajan, and R.L. Blackburn. *Algorithmic and Register-Transfer Level Synthesis: The System Architect's Workbench.* Kluwer Academic Publishers, 1990.

[179] Donald E. Thomas. *The Design and Analysis of an Automated Design Style Selector.* PhD thesis, Department of Electrical Engineering, Carnegie-Mellon University, 1977.

[180] Donald E. Thomas and Philip R. Moorby. *The Verilog Hardware Description Language.* Kluwer Academic Publishers, June 2002.

[181] C. J. Tseng and D. P. Siewiorek. Automated synthesis of data paths on digital systems. *IEEE Transactions on Computer-Aided Design of Integrated Circuits and Systems*, 5(3):379–395, July 1986.

[182] Underbit Technologies Inc. MAD: MPEG audio decoder. http://www.underbit.com/, 2008.

[183] University of California, Los Angeles (UCLA). The xPilot System. http://cadlab.cs.ucla.edu/soc/, 2008.

[184] Frank Vahid and Tony Givargis. *Embedded System Design: A Unified Hardware/Software Introduction.* John Wiley and Sons, October 2001.

[185] Frank Vahid, Sanjiv Narayan, and Daniel D. Gajski. SpecCharts: A VHDL front-end for embedded systems. *IEEE Transactions on Computer-Aided Design of Integrated Circuits and Systems*, 14(6):694–706, June 1995.

[186] J. Vanhoof, K. V. Rompaey, I. Bolsens, G. Goossens, and H. DeMan. *High-Level Synthesis for Real-Time Digital Signal Processing.* Kluwer Academic Publishers, 1993.

[187] VaST Systems Technology Corporation. http://www.vastsystems.com/.

[188] Diederik Verkest, Karl Van Rompaey, Ivo Bolsens, and Hugo De Man. CoWare: A Design Environment for Heterogeneous Hardware/Software Systems. *Design Automation for Embedded Systems*, 1(4):357–386, October 1996.

[189] Virtutech. Virtutech Simics. http://www.virtutech.com/.

[190] K. Wakabayashi and T. Yoshimura. A resource sharing and control synthesis method for conditional branches. In *International Conference on Computer Aided Design*, pages 62–65, November 1989.

[191] Kazutoshi Wakabayashi. Cyber: High level synthesis system from software into asic. In Camposano and Wolf, editors, *High-Level Synthesis*. Kluwer Academic Publishers, 1991.

[192] John Waldron. *Introduction to RISC Assembly Language Programming.* Addison-Wesley, 1998.

[193] Andy Wellings. *Concurrent and Real-Time Programming in Java.* Wiley, 2004.

[194] Reinhard Wilhelm, Jakob Engblom, Andreas Ermedahl, Niklas Holsti, Stephan Thesing, David Whalley, Guillem Bernat, Christian Ferdinand, Reinhold Heckmann, Tulika Mitra, Frank Mueller, Isabelle Puaut, Peter Puschner, Jan Staschulat, and Per Stenström. The worst-case execution time problem: Overview of methods and survey of tools. *ACM Transactions on Embedded Computing Systems (TECS)*, 7(3):1–53, April 2008.

[195] Wayne Wolf. *Computers as Components.* Morgan Kaufmann, 2001.

[196] Xilinx Inc. FPGA and CPLD Solutions from Xilinx, Inc. http://www.xilinx.com/, 2008.

[197] Haobo Yu. *Software Synthesis for System-on-Chip*. PhD thesis, Information and Computer Science, University of California, Irvine, June 2005.

[198] Henning Zabel, Wolfgang Mueller, and Andreas Gerstlauer. Accurate RTOS modeling and analysis with SystemC. In Wolfgang Ecker, Wolfgang Mueller, and Rainer Doemer, editors, *Hardware Dependent Software: Principles and Practice*. Springer, 2009.

[199] Gerhard Zimmermann. The MIMOLA design system: A computer aided digital processor design method. In *Design Automation Conference*, pages 53–58, San Diego, CA, United States, June 1979.

Index